T0135194

Intelligent Systems Reference Library

Volume 55

Series Editors

Janusz Kacprzyk, Polish Academy of Sciences, Warsaw, Poland
e-mail: kacprzyk@ibspan.waw.pl

Lakhmi C. Jain, University of Canberra, Canberra, Australia
e-mail: Lakhmi.Jain@unisa.edu.au

For further volumes:
http://www.springer.com/series/8578

About this Series

The aim of this series is to publish a Reference Library, including novel advances and developments in all aspects of Intelligent Systems in an easily accessible and well structured form. The series includes reference works, handbooks, compendia, textbooks, well-structured monographs, dictionaries, and encyclopedias. It contains well integrated knowledge and current information in the field of Intelligent Systems. The series covers the theory, applications, and design methods of Intelligent Systems. Virtually all disciplines such as engineering, computer science, avionics, business, e-commerce, environment, healthcare, physics and life science are included.

Manuel Mora · Jorge Marx Gómez
Leonardo Garrido · Francisco Cervantes Pérez
Editors

Engineering and Management of IT-based Service Systems

An Intelligent Decision-Making Support Systems Approach

Editors
Manuel Mora
Department of Information Systems
Autonomous University of Aguascalientes
Aguascalientes
Mexico

Jorge Marx Gómez
Department of Computing Science
Carl von Ossietzky University of Oldenburg
Oldenburg
Germany

Leonardo Garrido
Departamento de Ciencias
Computacionales
Tecnológico de Monterrey
Monterrey
Mexico

Francisco Cervantes Pérez
CCADET
Universidad Nacional Autónoma de México
Ciudad Universitaria
Mexico

ISSN 1868-4394
ISSN 1868-4408 (electronic)
ISBN 978-3-662-51037-7
ISBN 978-3-642-39928-2 (eBook)
DOI 10.1007/978-3-642-39928-2
Springer Heidelberg New York Dordrecht London

Softcover reprint of the hardcover 1st edition 2014

Printed on acid-free paper

Springer is part of Springer Science+Business Media (www.springer.com)

Preface

A service economy has been recognized as the dominant paradigm in present times [1]. Such a service-oriented worldview demands new engineering and management scientific (both fundamental and applied) knowledge to cope with the planning, design, building, operation and evaluation, and disposal of non-adequate IT-based service systems [2]. Such challenges emerge from the paradigm shift from a product-based manufacturing economy to this new service-oriented one [3].

In turn, Intelligent Decision-Making Support Systems (i-DMSS) are specialized IT-based systems that support some or several phases of the individual, team, organizational, or interorganizational decision-making process by deploying some or several intelligent mechanisms [4, 5]. In particular, Artificial Intelligence (AI) has been recognized as a significant enhancement tool for DMSS [6, 7] since several decades. However, the utilization of i-DMSS for engineering and management of IT-based service systems is still scarce.

We believe that fostering its research and utilization is relevant and needed for advancing the progress of IT-service systems. Consequently, in this book we will pursue to following academic aims: (i) generate a compendium of quality theoretical and applied contributions in Intelligent Decision-Making Support Systems (i-DMSS) for engineering and management IT-based service systems (ITSS); (ii) diffuse scarce knowledge about foundations, architectures, and effective, and efficient methods and strategies for successfully planning, designing, building, operating, and evaluating i-DMSS for ITSS, and (iii) create an awareness of, and a bridge between ITSS and i-DMSS academicians and practitioners in the current complex and dynamic engineering and management ITSS organizations [8].

In this book, we have collected 11 chapters addressing such aims from international research community. These 11 chapters refer to relevant topics for both IT service systems and i-DMSS including: problems of selection of IT service providers, optimization of supply chain systems, IT governance decisions, clinical decision support, dynamic user-interface adaptation, re-engineering of processes, IT BSC decisions, and generic decision problems. Advanced IT technologies used in some chapters are: fuzzy multi-criteria mechanisms, semantic processing, data mining processing, and rough sets. Other chapters report traditional DSS mechanisms but used or suggested to be used in innovative mode for IT service engineering and management tasks.

In the first chapter entitled *Knowledge Management in ITSM: Applying the DIKW Model*, Sue Conger in University of Dallas, USA and Jack Probst, in Pink Elephant Company, USA, discuss the foundations of the data-information-knowledge-wisdom (DIKW) conceptual model and elaborate its usage justification in ITSM key activities. In particular, the DIKW model effects on the ITSM Service Desk activity are evaluated and analyzed. The authors report a case study where the utilization of a KMS and the DIKW model provides benefits from a tribal and trivial knowledge usage to advanced knowledge patterns plausible to become in wisdom chunks. This chapter, thus, contributes for both ITSM and iDMSS topics providing how KMS and DIKW model can support and enhance ITSM activities and decisions.

In the second chapter entitled *IT Governance Practice in a Malaysian Public Institute of Higher Learning and Intelligent Decision Making Support System Solution*, Abdul Rahman, Yusri Arshad, and Binyamin Ajayi, all of them in International Islamic University Malaysia, in Malaysia, present a case study conducted in a Malaysian university on their institutional efforts for introducing and using IT governance mechanisms. Authors indicate that better IT governance practices lead usually to a better utilization of IT resources (financial, technical, human, organizational, informative), which lately should impact on the university's performance. COBIT is used a model of reference. After it, authors identify the strong need to support IT governance mechanism implementation with iDMSS. Two examples of plausible utilization are reported. This chapter contributes with an empirical study on the problematic of IT governance mechanism implementation in an academic setting, and how it can be supported using iDMSS.

In the third chapter entitled *DSS-based IT Service Support Process Reengineering using ITIL: A Case Study*, Raul Valverde and Malleswara Talla, both in Concordia University, Canada, present a case study where a reengineering of ITSM processes was conducted. The authors report that while exist several ITSM frameworks, most of them report only general recommendations. Thus, companies need to invest additional resources for deploying a particular process. In this case study, the company elaborated a traditional decision support system (DSS) for tracking a set of key performance indicators (KPIs) associated to several ITSM processes. By using such traditional DSS, the company was able to reengineering their processes and obtain improved ITSM metrics. This chapter, despite does not present an iDMSS, contributes to advance our knowledge on ITSM best practices introducing DSS for tracking KPIs that lately guide and justify ITSM reengineering efforts.

In the fourth chapter entitled *Managing Cloud Services with IT Service Management Practices*, Koray Erek, Thorsten Pröh, and Rüdiger Zarnekow in Technical University of Berlin, elaborate the case of the relevance of cloud computing for modern ITSM practices. The authors point out the importance of utilization of ITSM best practices in organizations, as well as the explosive growth of cloud computing service providers. However, the interrelationships and effects of the latter (cloud computing) on the former (ITSM practices) has been scarcely studied. Thus, the authors discuss and analyze the implications of hiring IT services from cloud computing model on the internal ITSM practices. They suggest that such implications should be considered when the organization relies on distributed services in order to obtain an effective management. This chapter strictly does not report a particular application of iDMSS to ITSM. Nevertheless, this chapter presents a core topic in services (cloud computing services) and elaborates a map of current ITSM tools which could be enhanced with iDMSS capabilities.

In the fifth chapter entitled *Towards Semantic Support for Business Process Integration*, Roberto Pérez and Inty Sáez, in Central University "Marta Abreu" of Las Villas, Cuba, and Jorge Marx-Gómez in Carl von Ossietzky University of Oldenburg, Germany, present an analysis of the major contributions of the semantic processing approach to the business process integration. The authors also report a suggested design framework and its illustration with supply chain management processes. This chapter does not address directly the ITSM theme. However, from a software engineering view, the integration of process in workflow and business process management systems (BPMS) is a relevant theme. Furthermore, such problematic implies both the deployment of services and advanced IT mechanisms (like semantic processing). Thus, we believe this chapter can be useful to foster-related research in the specific themes of ITSM and iDMSS.

In the sixth chapter entitled *Integrating ERP with Negotiation Tools in Supply Chain*, Paolo Renna in Università degli Studi della Basilicata, Italy, report a new architecture of a multi-agent system for supply chain negotiations and its simulation results. In particular, this study refers to the Build to Order (BTO) operation mode where no previous stock exists of products. Thus, negotiations for getting timely and cost-effective components in a buyer centric e-marketplace become critical. Similar to the previous chapter, this does not address directly the ITSM theme. However, being ERP and SCS two of the most used IT services, their improvements are relevant for IT service implementers. Thus, we believe this chapter can be useful to foster related research in the specific themes of ITSM and iDMSS.

In the seventh chapter entitled *Adaptive Applications: Definition and Usability in IT-based Service Systems Management*, Ammar Memari and Jorge Marx-Gomez, both in Carl von Ossietzky University of Oldenburg, Germany, study the notion of adaptive application. First, the authors elaborate an informal definition from the literature review. Second, by using rough sets, the authors elaborate a formal definition. Finally, authors illustrate its usefulness in the field of IT service management. This chapter contributes directly to both ITSM and iDMSS research streams introducing the description power of rough sets, and how this mechanism can be used to formalize relevant concepts in ITSM. Such foundations can be further incorporated in a particular iDMSS.

In the eighth chapter entitled *Attitude-based Consensus Model for Heterogeneous Multi-criteria Large-Scale Group Decision Making: Application to IT-based Services Management*, Ivan Palomares and Luis Martinez in University of Jaen, Spain, present an innovative mechanism for reaching consensus in large decision groups: a fuzzy multi-criteria attitude-based model. The authors describe the characteristics, concepts, and algorithms of this mechanism and illustrate its usefulness with a real case of selection of IT services in the banking domain. This chapter contributes directly and strongly to both ITSM and iDMSS research streams. Group decision-making is a relevant topic scarcely studied and applied in ITSM at present. Thus, this chapter helps to promote and foster the research and utilization of advanced decision-making practices in ITSM.

In the ninth chapter entitled *Improving Decision-Making for Clinical Research and Health Administration*, Alexandra Pomares and Rafael Gonzalez in Javeriana University, and Wilson Bohórquez, Oscar Muñoz, Dario Londoño, and Milena García, in San Ignacio University Hospital, in Colombia, report the experience gained in the development and implementation of a health decision-support system called DISEArch using a service approach. DISEArch uses data mining mechanisms for extracting useful knowledge from EHR (electronic health records). Initial empirical results show clear benefits of the iDMSS on the provided IT health service in the real setting. This chapter, thus, contributes to ITSM and iDMSS—from a software engineering view—with a report on the knowledge and learned lessons of an advanced IT service.

In the tenth chapter entitled *Architecture for business intelligence design on the IT service management scope*, Pablo Marin-Ortega and Patricia Pérez-Lorences in Central University of Las Villas, Cuba, and Jorge Marx-Gómez in Carl von Ossietzky University Oldenburg, Germany, develops a working prototype of an intelligent DMSS for supporting IT BSC decisions and metrics. For its design authors use Principal Component Analysis statistical method and Compensatory Fuzzy Logic mechanisms. Authors illustrate the potential usefulness of this intelligent DMSS with data from an IT area in a real business setting.

In the last eleventh chapter entitled *Improving IT Service Management with Decision-Making Support Systems*, Manuel Mora in Autonomous University of Aguascalientes, Mexico, Gloria Phillips-Wren in Loyola University, USA, Francisco Cervantes-Pérez in CCADET UNAM México, Leonardo Garrido in Monterrey Tech, México, and Ovsei Gelman, in CCADET UNAM México, elaborate the case on the need of incorporating traditional and advanced decisional support through decision support systems (DSS) and intelligent DSS in all of the ITSM processes. Firstly, authors describe the foundations of decision-making process and DSS and i-DMSS. Secondly, using ITIL v3 a reference model, they describe the main ITSM processes and their implicit and explicit decisional situations. Finally, authors illustrate two examples (one for DSS and the other for i-DMSS) on how two specific ITSM decision situations can be supported by such systems. Lately, authors suggest that your research can be useful to foster more specific research on DSS and i-DMSS for ITSM.

Hence, we believe that these 11 chapters fit the aims of this book either directly or indirectly. However, we are sure that all of them report relevant and rigorous research useful for ITSM and iDMSS research streams and they are worthy to be disseminated in our research communities. Thus, we thank our colleagues who have contributed to the realization of this book submitting a high quality chapter, helping with the internal asked reviews, and improving their accepted chapters using the reviewer's recommendations. These colleagues represent an adequate international landscape: USA, Germany, Canada, Italy, Spain, Mexico, Colombia, Malaysia, and Cuba.

We also thank Prof. Dr. Janusz Kacprzyk of the Polish Academy of Sciences, Systems Research Institute, Poland, and to Prof. Dr. Lakhmi C. Jain, Founding Director of the KES Centre, University of South Australia, Australia, editors of the Springer Intelligent Systems Reference Library book series, and the Springer staff for all the provided support for this book. Finally, we thank respectively our working institutions: Autonomous University of Aguascalientes, Mexico; Oldenburg University, Germany; Tecnológico de Monterrey, México; and CCADET Universidad Nacional Autónoma de México, México; for their continual support for our research activities.

Aguascalientes, Mexico — Manuel Mora
Oldenburg, Germany — Jorge Marx Gómez
Monterrey, México — Leonardo Garrido
Ciudad Universitaria, México — Francisco Cervantes Pérez

References

1. Chesbrough, H., Spohrer, J.: A research manifesto for services science. Commun. ACM **49**(7), 35–40 (2006)
2. IfM and IBM. Succeeding Through Service Innovation: Developing a Service Perspective for Education, Research, Business and Government. University of Cambridge Institute for Manufacturing, Cambridge (2008)
3. Dermikan, H., Spohrer, J., Krishna, V.: Introduction of the science of service systems. In: Dermikan, H., Spohrer, J., rishna, V. (eds.) The Science of Service Systems. Service Science: Research and Innovations in the Service Economy Series, 1–10, Springer, New York (2011)
4. Forgionne, G.A., Gupta, J.N.D., Mora, M.: Decision making support systems: achievements, challenges and opportunities. In: Mora, M., Forgionne, G., Gupta, J.N.D. (eds.) Decision Making Support Systems: Achievements and Challenges for the New Decade, pp. 392–403. Idea Group, Hershey (2002)
5. Phillips-Wren, G., Mora, M., Forgionne, G., and Gupta, J.: An integrative evaluation framework for intelligent decision support systems. Eur. J. Oper. Res., **195**(3), 642–652 (2009)
6. Goul, M., Henderson, J., Tonge, F.: The emergence of artificial intelligence as a reference discipline for decision support systems research. Decis. Sci., **23**, 1263–1276 (1992)
7. Eom, S.: An overview of contributions to the decision support systems area from artificial intelligence. Proceedings of the AIS Conference, Baltimore, 14–16 Aug (1998)
8. Mora, M., O'Connor, R., Raisinghani, M., Macias-Luevano, J., and Gelman, O.: An IT service engineering and management framework. Int. J. Serv. Sci. Manage. Eng. Tech. **2**(2), 1–16 (2011)

Contents

Chapter 1
Knowledge Management in ITSM: Applying the DIKW Model

Sue Conger and Jack Probst

Abstract The data-information-knowledge-wisdom (DIKW) model is known as applying to information processing concepts. Similarly, information technology service management (ITSM) has enjoyed rapid industry adoption as a set of processes for managing IT infrastructure and the delivery of IT services. However, data generated during and supporting ITSM activities often stays at the information stage. This paper evaluates the DIKW model and its application to ITSM because the information, knowledge, and wisdom levels most effectively guide organizational improvement. Thus, to gain the maximum benefit from ITSM, companies must strive to record created information, facilitate knowledge development, and identify the individuals in the organization who can apply wisdom to particular complex challenges. The paper first discusses the DIKW model, then outlines key activities of ITSM that are amenable to DIKW application. Next, use of DIKW for improving the conduct of ITSM Service Desk activity is evaluated and discussed. Typical Service Desks stop at support of work with a known errors database, automating at a knowledge level of work. A case study shows how an incident knowledge management database to automate outages and decision patterns, moving toward the wisdom level, can move beyond tribal knowledge to institutionalize knowledge patterns, resulting in reduced costs and faster incident resolution.

Keywords DIKW model · IT service management · ITSM service desk · ITIL · KM · IKMDB

S. Conger (✉)
University of Dallas, TX 75062, USA
e-mail: sconger@utdallas.edu

J. Probst
Pink Elephant, IL 60008, USA
e-mail: jprobst@pinkelephant.com

M. Mora et al. (eds.), *Engineering and Management of IT-based Service Systems*,
Intelligent Systems Reference Library 55, DOI: 10.1007/978-3-642-39928-2_1,

1 Introduction

The purpose of this research is to introduce a new way of thinking from solutions to problems to allow identification and integration of solution fragments to generate new solutions to Service Desk problems and, thus, improve the FCR to closer to 100 %. This research is important because the higher the FCR, the lower the cost of service, the higher the customer satisfaction, and the happier the service staff. In addition, the research is important because Service Desk work is representative of many types of discretionary knowledge work that might profit from similar treatment.

The state of knowledge management (KM) has advanced to low-level, repetitive tasks but has eluded highly complex or collaborative tasks [6]. The status of KM provides opportunities for automation in many types of knowledge work but relies primarily on the expertise and insight of the knowledge worker performing the work to translate information to knowledge [9]. Information technology (IT) is well understood for automation of low level KM, however intelligent automated support for many KM tasks has been elusive [6].

Much IT knowledge work relates to the provision of services to the parent or client organization. IT service management (ITSM) encompasses a body of best practices for instance, the ITIL framework [38], for how IT creates or delivers value to the business through provisioning and delivery of services. Service Management frameworks such as ITIL articulate the IT processes that set the stage for services provision. After first optimizing specific IT processes, an organization embeds the service in a governance and management structure, redefining work, job definitions, and pay structures to encourage services provision. However, regardless of adopting frameworks, such as ITIL, many organizations' services are not sufficiently structured to profit from them, regardless of the other actions taken to ensure quality service provision [20]. Rather, unstructured IT services require ingenuity and improvisation by the service professional in both determining the service required and how it is delivered. As a result, simply following a framework, such as ITIL, for ITSM is insufficient to guarantee quality service delivery [32]. One example of a service area that illustrates these issues is the Service (or help) Desk. A Service Desk provides service support for incidents that are some type of IT failure, requests (e.g., password reset), services to access protected or auditable data, or events that encompass monitoring for potential IT failures [38]. However, a disorganized or unstructured approach to Service Desk management can impede service delivery. Without the structure of information capture, search, and reuse, the service desk function can be limited to simple call center operations.

A systems approach to Service Desk services requires evaluation of the tasks based on input, process, output, and feedback mechanisms [39]. The inputs are issues, outages, requests, or events. Processes include logging tickets, search of a known errors database (KEDB), troubleshooting, conversing with the client, possible discussions with vendors, and escalating if no solution is found. The output is a successful resolution to the request. Feedback to regulate the process takes the form of metrics on Service Desk operations, including first contact resolution (FCR), customer satisfaction, and so on. Typically, Service Desks strive for 85 % or higher FCR that often relies on finding a KEDB entry matching the problem. When no KEDB resolution is found, the process relies on tacit knowledge, that is, personal expertise of the individual Service Desk employee.

KEDBs are databases that summarize issues, outages, and their resolutions or workarounds [38]. A KEDB search is based on the issue or outage and, once a match is found, the solution is given to the user to test that the issue is resolved. The challenge to a useful KEDB is to develop search terms that allow a variety of ways to find the information about the problem. Searches move from issue to solution, thus, the issue must be defined properly to support finding the solution. Typically, Service Desk employees are taught this linear issue-to-solution method of thinking to facilitate use of the KEDB and their own personal troubleshooting. With this way of thinking, FCRs are constrained to something less than 100 %, in the best cases, around 95 %. This research seeks to develop a method of KM that increases this percentage.

2 Foundations of Knowledge Management and IT Service Management

2.1 Knowledge Management

A knowledge management (KM) is the systematic process of acquisition, organization, and communication of organizational knowledge for reuse by others in the community [3]. Knowledge relates to an individual's ability to take an action and can relate to declarative (know-what), procedural (know-how), causal (know-why), conditional (know-when), or relational (know-with) types of knowledge (Zack, 1998 as cited in [3]). Each knowledge type needs to be examined to determine the extent to which the project being documented needs the type of information and the extent to which the information can be codified. Some organizations only code the fact of the information and identify the person to contact, providing a directory of subject matter experts for their knowledge repository [3]. Recording of both searchable information and a directory of information sources are considered best practices.

Further, knowledge is categorized as tacit or explicit. Tacit knowledge is in an individual's mind and relates to the individual's experiences with technical aspects in terms of skills or procedural knowledge (Nonaka, [28]). Explicit knowledge is known to an organization, codified, and shareable (Nonaka, [28]). Tacit knowledge poses the larger challenge to KM as expertise and reasoning processes are difficult to identify clearly [6, 11, 24, 28]. Professionals are often not able to articulate their reasoning processes. Plus, when solutions are constantly tailored to situations, pre-defining situational needs may not be practical or possible [18]. Explicit knowledge on the other hand, is simply the codification of relatively well-understood knowledge into a reusable format [33].

2.2 Data-Information-Knowledge-Wisdom

One model that can facilitate KM in a Service Desk relates to the development of wisdom, starting at data, progressing to information, then knowledge and wisdom [1, 2]. Data are the raw material, numbers and letters that describe some phenomena [25]. In organizations, data are generated as discrete units of measurement. Numbers or letters without a context or reference remain just numbers and letters [25]. By applying a referential perspective, data becomes information. For example, numbers such as 123456789 formatted as 123-45-6789 becomes a social security number that uniquely identifies an individual in the U.S. Thus, information is data with meaning, providing a lens for interpretation.

Knowledge is the expertise and skills acquired through experience and education; knowledge is the understanding of information and information patterns within a context [18, 27]. By itself, for instance, a social security number has limited meaning but if it is within the context of a tax return that links to the patterns of an individual's financial history, it provides knowledge. Knowledge is supported by the flow of information between individuals and develops through experience [27]. Information patterns or knowledge can be explicit, codified, and recorded or they remain tacit to individuals [28].

The value of knowledge derives from information use in analysis, decision-making, problem solving, and teaching [18]. Application of knowledge allows us to recognize situations that are similar to past situational patterns through remembered, stored, compared, and retrieved information [27, 28], Nonaka and van Krogh, [29]. Patterns are key to knowledge transfer. Recognition of previously analyzed information patterns provides the basis for some future action or decision and provides greater context for the use of information [7, 15, 28].

Wisdom is defined as the experience-based capability to derive a deep and profound level of understanding of patterns or key relationships critical to an activity [28]. A pattern is an exemplar or repetitive structure based on characteristics, outcomes, or actions of some setting. With wisdom, one is able to discern complex patterns and to apply those in ways that are unique, creative, or ground breaking. Thus, wisdom extends knowledge but requires a higher level of cognitive processing to discern, rationalize, analyze, assemble, or reconfigure patterns in ways that have not been explored or are not typically obvious. Elkhonon (Goldberg [14]) summarizes the concept of wisdom as "the ability to connect the new with the old, to apply prior experience to solution of a new problem… a keen understanding of what action needs to be taken." Because wisdom leads to a new contextual characterization or new outcome, it is difficult to codify as a digital asset. 'Wisdom workers' are a small set of individuals whose knowledge and expertise are desired to be codified by organizations. In organizations, wisdom workers are a few key individuals who possess the most experience, who are innovative, and who are most skilled in problem analysis and problem-solving [14].

In an organizational setting, knowledge is the province of most individuals during the normal course of their jobs. They develop a referential experience base that they apply to repeated situations (Davenport, et al., 1998, [13, 21]). Knowledge workers call upon this experience base to solve challenges. The organizational challenge is to take data gathered from process conduct, digitize patterns of solutions to create information, and then learn to transform information into knowledge in some way, eventually institutionalizing, internalizing and automating the learned experience in job aids applicable to future work [23]. This challenge has mostly been solved [18]. The next step is to analyze knowledge to determine new ways to combine the information to create wisdom [6]. Most organizations are successful at providing digital information aids; few are successful at automating problem elements that are combinable into unique ways to create new solutions to problems [6].

2.3 IT Service Management

IT Service Management, servitizing the IT function, has emerged as a strategic focus for industry that furthers business—IT alignment goals [4, 38]. A simple service refers to organizational capabilities for providing value to customers in the form of an experience that is consumed as it is produced [30, 41]. An IT service situates defined processes within a governance and management structure, defines number and nature of work for multiple locations, defines level of discretion in the work, defines software, data, and IT resource support for the functions and roles, and defines service levels for customer delivery including response time, service desk response support, metrics, and so on [5]. Thus, service design differs from typical application design by defining not only the application but also the organizational context, supporting IT infrastructure, IT support structure, and metrics, which all are then integrated to standardize customer interaction and provide customer (business) value [5].

The DIKW model applies to ITSM because ITSM encompasses knowledge work in IT. Significant data are generated in the course of everyday IT operations and service delivery. Data is often in the form of metrics but also may be generated in the course of conducting business, such as the research involved in answering a novel question by the Service Desk. Data, after initial generation, can become information when analyzed, contextualized, and presented in a dashboard or report. This data information can become knowledge when analyzed and used to inform decisions. Finally, this knowledge can become wisdom when the unique experiences of individuals are applied to the information with insight to solve difficult challenges or develop unique solutions. The problem is that the data can also be lost in the course of business conduct as the same data are generated multiple times on multiple occasions when KM is not practiced. Without an organizational capability to structurally capture and categorize data in such a way that supports reuse, valuable experiences are lost. Efficient delivery of services relies upon organized use of scarce resources, not the least of which is how service workers reuse past experience to solve the ongoing challenges of the service environment. Thus, some mechanism for turning data into information into knowledge into wisdom is a desired outcome for Service Desk success.

This reasoning allows one to ignore criticisms of DIKW model and its development over time in an organization, which criticizes linear development of wisdom and ignores exogenous influences such as culture on the process [10]. Use of the model simply recognizes that the four elements—data, information, knowledge, and wisdom, differ and build on each other. There is no necessary distinction about how they develop, which could be linear, skipping one or more steps, or due to exogenous influences [10]. Knowledge is fairly well understood in the Service Desk KEDB context.

Both tacit and explicit knowledge occur in Service Desk work. Explicit knowledge of past computer outages, is codified into a 'known errors database' (KEDB), which is a searchable, online database accessible by all Service Desk support staff [38]. When an outage occurs and is recorded, the KEDB is checked to determine if a solution or workaround already exists. When found, the solution is applied and verified as working.

When no solution is found, the expertise of the Service Desk staff in the form of tacit knowledge takes over. Then, the worker reviews his past experience to determine similar situations and tries to resolve the issue based on that past experience (Van Grich and Soloway, [39]). Alterations and improvised solutions based on experience may result. Ideally, then, the new solution should be able to be codified and added to the KEDB but often, the circumstances are so unique that the solution is unlikely to ever be used again in exactly the same way.

An example might be an outage due to a vendor engineering change. The reasoning used to find the root cause is interesting but the solution is mundane, rarely applicable, and not worth the effort to put into a KEDB. Separating those tacit situations of interest from those that are not is a subjective, difficult prospect [18]. Further, the reasoning may be more important than either the problem or the individual solution but documenting that in a KEDB is not usually feasible. Some method of capturing reasoning and solutions without necessarily tying them to an individual issue is desirable to provide closer to 100 % automated coverage of Service Desk issues (cf. [6]). Therefore, the challenge is to support development of wisdom workers such that all Service Desk staff know the techniques and principles and are encouraged to develop 'wise' solutions to previously unknown issues. Challenges to this process are in identifying tacit knowledge of interest and supporting development of novel solutions from prior solution and reasoning knowledge.

3 Applying DIKW to ITSM

3.1 *Knowledge Management Application*

The objective of KM is to enable organizations to improve the quality of management decision making and to support the execution of future tasks or decisions by ensuring that reliable and relevant information is available throughout the services lifecycle and that organization learning from past activities is not lost. To sustain an organizational approach, a KM strategy is needed that encompasses the span of KM activities from sense making, to identification and acquisition, to cataloging and storage, to knowledge retrieval and transfer [19]. These concepts are simple in theory but not in practice. Some challenges to KM, and therefore to moving from information to knowledge, are differences in learning styles, culture interpretation and presentation, and design of appropriate architectures and acquisition practices [18, 19]. Moving from knowledge to wisdom requires recognition and association of unique patterns, contexts, and relationships and how those configurations might be combined to form new insights and information.

Service delivery encompasses activities for IT operations and production. In the context of ITSM, information support requires means to not only internalize information and knowledge for organizational use, but also to alter the way processes are conducted as a result of previously documented organizational decisions, analysis or problem-solving patterns. ITSM defines a set of structured processes and organization design to foster repeatable organizational behaviors. These processes address the management of operations, risk and the service environment. As one might expect, the service environment generates repetitive situations or activities. Thus, a KM strategy is a valuable ITSM mechanism to efficiently and effectively address the repeatable nature of the service environment. To do this, KM practices in the form of pattern recognition, retrieval, and automation are needed.

Data gleaned from IT service conduct, often in the form of metrics, can be analyzed from the perspective of the process under scrutiny and aggregated to develop recognizable patterns. Patterns are used to recognize, abstract, group, and analyze data (De Bono, [8]). Patterns are important as a tool for the knowledge worker or subject matter expert (SME) to leverage experience and avoid the pitfall of relearning the same or similar experiences. When the brain recognizes a pattern from the past (such as a similar type of service problem), it ignores the onslaught of additional, potentially confusing, information, which obfuscates the challenge at hand. That is, the person 'stops thinking' in order to analyze a particular previous solution or solutions associated with the problem pattern at hand (De Bono, [8]). The cognitive process is that the previous pattern-solution set is analyzed, matched or modified as needed to fit the current context and then is associated as the solution for the challenge at hand (De Bono, [8]). In the event that a modification of pattern-solution set is required, the modified pattern-solution is remembered as a new pattern for use in future problem solving searches.

Pattern recognition is typical of how organizations 'learn' [11, 33]. To illustrate this point, reflect on how one learns algebra. Students learning algebra are taught to identify specific problem and equation 'patterns' and, once recognized, how to set up a specific solution pattern to solve the problem. For instance, one starts the learning process with the mechanics of simple equations and relationships, such as formulae for linear equations and how to solve them. An example is $2x = 12$. The patterns of $ax = c$ requires one to divide c by a to obtain x. In applying this knowledge to a slightly more complex problem, one uses the patter. For instance, the problem $2x + 4 = 12$ exemplifies the next step. The pattern of $ax + b = c$ and how to solve for x are two patterns recognized as leading to the solution: change the problem terms by subtracting four from both sides of the equation to get $2x = 8$; then divide by 2 to get $x = 4$. Students learn to recognize the pattern and how to apply a predefined solution methodology to the recognized pattern. Pattern recognition continues through more complex problems and solutions, adding to a student's skills or knowledge base. Students progress until they understand how every polynomial can be factored into first and second-degree polynomials using a combination of pattern recognition and skill [15].

In the example of algebraic formulae, pattern recognition leads to recognition of solution pattern sets based on factoring, the solution technique [15]. The knowledge (or learning experience) for complex equations relies on prior knowledge gained in pattern recognition and problem solving for simple equations, building skills through pattern recognition and application of prior skills. As knowledge develops, learning progresses to problems with two variables, three variables, and so on. The point is that as students acquire new skills, they build upon and rely on all prior skills ([15, [16]). Thus, as learning progresses, knowledge of techniques associated with patterns are identified, codified, stored, sequenced, and retrieved for reuse in many situations ([15, 33]). The student learns how to identify defined patterns, retrieve previously learned pattern-solution sets, and apply them to the problem at hand [15].

ITSM requires similar pattern recognition and problem solving. A Service Desk example illustrates the KM challenges. The Service Desk provides services to resolve incidents (or outages), questions about IT usage, requests relating to IT usage, and data access requests. Patterns of these problem types recur daily in Service Desk operation. Thus, a database failure in one context may have the same solution as a database failure in another context, with context identifying the application, server, network, or other infrastructure item. An ITSM discipline for identifying, defining and managing pattern-solution sets is called knowledge-centered support' [38]. Knowledge-centered support defines a set of practices to create, reuse, and improve knowledge for use within the service delivery processes (Robertson, et al., [33, 34, 35, 37, 42]). Knowledge-centered support differs from other KM practices in that it relates to not only to the creation of digital assets for recognition and retrieval, but also to the development of a socialization process for developing a community of collaboration [12, 27, 37].

3.2 Reverse Engineering to Codify Problem-Solution Pairs

In ITSM, companies create a known errors database (KEDB) to document recurring incidents or problems and their solutions for both the experienced and inexperienced Service Desk customer support representatives (CSRs). When an incident is reported, the CSR asks about the incident using a general question, asking progressively more specific questions to better understand the incident. Once armed with complete incident description information, the CSR queries a KEDB to determine if the incident he is working has occurred before and whether there is a solution or work-around to repair the outage [38].

The KEDB is commonly the only knowledge support for a Service Desk. Yet, a KEDB is really only automating information. Knowledge occurs when the information is used in service of an issue. As a result, many issues occur and, as the KEDB is queried and no solution is found, individuals must rely on their own experience or their peers' tribal knowledge to determine the best course of action [24]. Service Desk staff apply their experience to identifying unfamiliar or previously unknown patterns of a problem to arrive at a solution. Once a solution is found, it is stored in the KEDB in the form of problem-solution knowledge for future retrieval and reuse.

Figure 1 depicts a typical search pattern and use of a KEDB followed by Service Desk, a customer support representatives (CSR). The pattern of search if problem-to-solution, or, front-to-back. The CSR asks questions, from general to increasingly specific to identify the class of problem, the context, the problem nature, and problem characteristics. Eventually, the CSR hones in on enough specific problem information to allow a search of the KEDB that will, hopefully, find the solution. Ideally, recognizing patterns and arriving at solutions could be codified to automate the search function. However, the number of potential factors underlying an incident can grow at an exponential rate if an answer is not relatively simple. This leads to a time-consuming resolution process or to no resolution found.

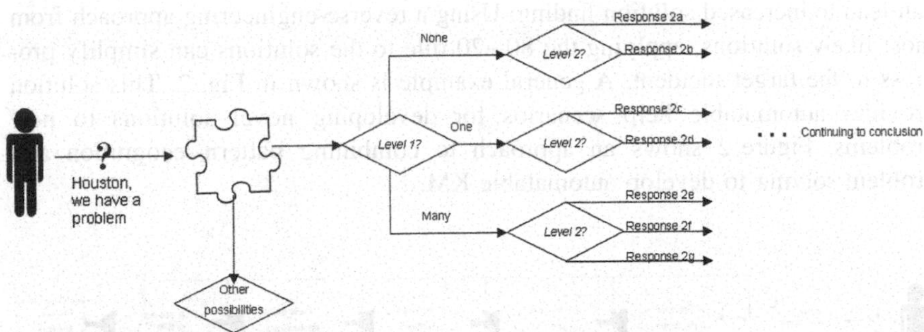

Fig. 1 Typical service desk search for incident-solution (Adapted from [31])

Problems can have four types of solution relationships: one problem to one solution (1:1), one problem to many solutions (1:N), many problems to one solution (N:1), or many problems to many solutions (N:M). KEDBs handle 1:1 and N:1 sets of problems and solutions well following the binary decision tree logic of a KEDB as shown in Fig. 1. However, for 1:N and N:M problem–solution sets, a KEDB appears not to be the most efficient resource. Using a KEDB for 1:N and N:M problems as depicted in Fig. 1 quickly becomes unwieldy. In addition, KEDBs typically contain causal know when and procedural know how information that identifies process and declarative issue characteristics; KEDBs do not document declarative know what or relational know with information, thus, limiting the potential for solution identification.

Thus, following KEDB logic, typically one would create a KM depiction of a problem by reasoning from problem-to-solution. However, when trying to identify patterns of problem-solution sets for 1:N or N:M problems, pattern recognition through a cause and effect analysis can be complex and, ultimately, not useful as it can still result in many potential solutions. This method of problem analysis has an exponentially growing number of qualifying conditions as one searches and does not find a solution. When a resolution is not in a KEDB, troubleshooting conducted by the Service Desk staff is currently the solution.

Troubleshooting is a general term meaning 'search for a solution.' It is a tacit-knowledge and experience-driven activity that varies widely in efficiency and effectiveness from one person to another and from one situation to another [36]. Similar to the learning of math patterns, experts assess a problem and tend to search by most likely solution sets versus novices who tend to search more randomly or, serially [15, 36].

A different approach, appears able to yield additional programmable outcomes by reverse engineering from likely solution to problem. Practical experience shows that for a typical Service Desk, both general issues and outage incidents often follow the 80–20 rule with 80 % of incidents resulting from 20 % of the causes. Therefore, when faced with a problem with many sets of potential solutions, a right to left, solution-to-problem methodology may be faster and more successful. Further, including relationship and conditional information in the searched data can lead to increased solution finding. Using a reverse-engineering approach from most likely solutions, applying the 80–20 rule to the solutions can simplify progress to the target incident. A general example is shown in Fig. 2. This solution provides automatable help scenarios for developing novel solutions to new problems. Figure 2 shows an approach to combining pattern-recognition and problem solving to develop automatable KM.

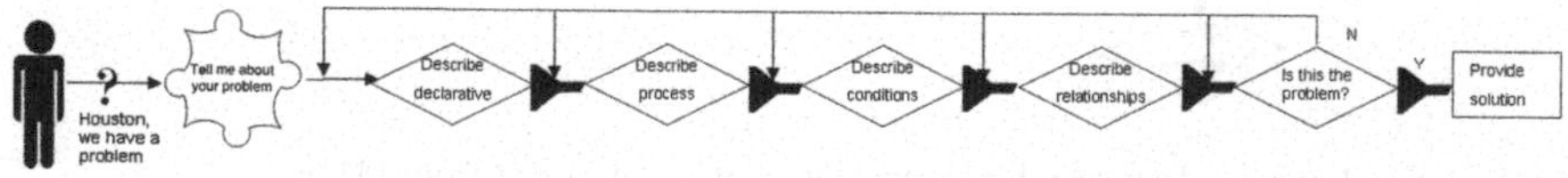

Fig. 2 Reverse engineering of incidents to obtain most probable solution (Adapted from [31])

Figure 2 shows an approach to reverse engineering in order to obtain the minimal number of questions needed to find incident characteristics. An incident characteristic is some feature of an incident or a solution that contributes to its uniqueness and can include relational, conditional, causal, declarative, or procedural information. Order of questioning would be directed by the tacit knowledge of the CSR, and, they would continue with characteristics questions until they could provide a solution, not necessarily including all of the characteristic types. Thus, tacit knowledge still plays a role in this process, but it provides even novices with a technique to reason from solution-to-problem by reverse engineering the issue. The reasoning process provides a funnel-like narrowing of solutions that could fit the problem. Each answer both narrows the number of solutions and also determines the next type of question to be asked.

Automation of this solution finding technique would result is a new database populated with incident characteristics that are related to solutions. Thus, the KEDB (forward analysis) is supplemented with an Incident Knowledge Management Database (IKMDB, backward analysis), a repository of solution-incident characteristic pairs. Development of the IKMDB is more complex than the KEDB. Known errors in the KEDB identify individual incidents and their solutions. Complex situations in which errors have ripple effects, or for which there are many possible solutions, are not easily documented in a way that allows simple selection of a most appropriate solution. An IKMDB builds both complexity of description and solution-incident characteristic relationships, and also supports intelligent search for patterns that transcend the individual solutions. For instance, characteristics of incidents might be searchable and present a variety other likely related incident characteristics or solutions.

In developing the IKMDB, incidents and solutions are reduced to solution-incident characteristic pattern pairs. The process for developing the information for the IKMDB is summarized in Fig. 3. Starting with the solution, subject matter experts (SMEs) are interviewed to identify the patterns they used to decompose, analyze, categorize, ever-larger series of incident characteristic -solution set patterns to get to the solution. Then, the SME process is reverse engineered to start at a solution and use only the successful forward search actions to define solution-incident characteristic pairs, eventually arriving at the original problem statement. The pattern-recognition activity decomposes a solution into its incident characteristics, which can relate to both incidents and solutions. An incident characteristic is some feature of an incident or a solution that contributes to its uniqueness. An incident characteristic may be an infrastructure characteristics (e.g., software package), an incident characteristic (e.g., cloud addressing error), data

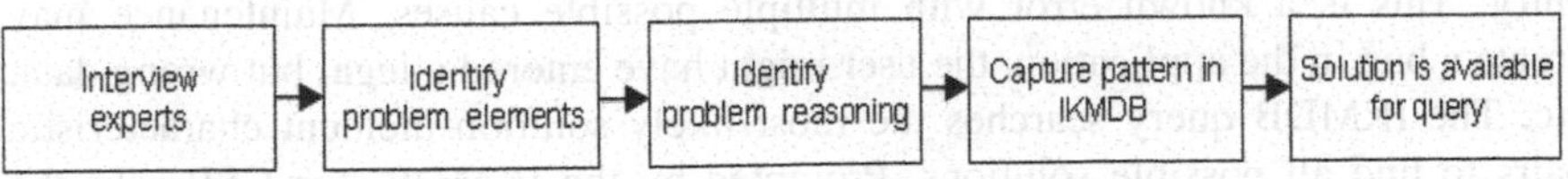

Fig. 3 Incident knowledge management database (IKMDB) original development process

characteristic (e.g., Gregorian date), or software characteristic (e.g., Python + XML), or other characteristic that provides part of a unique identification when a solution-incident characteristic set is aggregated. Incident characteristics that SMEs attend to (or ignore), and why, are identified and codified to be stored in the IKMDB. All users of the IKMDB then have access to solutions, the related incident characteristics, and the reasoning used to obtain them. Use of an IKMDB, plus the sessions analyzing solution-incident characteristics to build the IKMDB can build expertise across Service Desk staff.

When all known situations are identified and codified, the solution set of patterns is documented such that if a similar incident arises in the future, a defined patterned solution can be identified for reuse. When solutions suggested by the approach are not appropriate, a new solution pattern, through SMEs' application of wisdom is developed. Thus, patterns may not be recognizable as simply as single incident element-solution pairs but may require recognizing similarities across multiple incidents. The solution context is also coded to ensure correct pattern recognition. All of the incident, pattern, and solution information is automated in the IKMDB and forms the basis for future problem solving.

Issues to be considered in the IKMDB development process include how to define the elements, how to retrieve the elements, and the number of possible alternative steps to describe. For instance, matching of incidents may be by context, by incident, by incident characteristic, by frequency, by expected solution, or other criteria such as geographic location. Typically, there will be more than one potential solution to an incident in the IKMDB. These situations must also be accommodated and identified through the IKMDB query for reporting to the CSR for evaluation and testing.

3.3 IKMDB in Use

No database of incident sets is ever complete. As a result, the IKMDB needs to become a living part of the process of novel incident solution finding. Figure 4 depicts the process of IKMDB use. The CSR assimilates details of the outage, developing an understanding of the incident and its context. The incident is decomposed into components, as needed. Each component is analyzed and defined. Next, a query to the KEDB is made to determine if the incident has single known solution. When the KEDB search fails to find a known solution, the CSR queries the IKMDB entering incident characteristics in queries to match solution-incident characteristic patterns. Simple incidents would be incidents that are found because the context matches. For example, order entry aborts during simple data entry. This is a known error with multiple possible causes. Maintenance may create a bug in the application, the user might have entered a legal but wrong data, etc. The IKMDB query searches the most likely solution-incident characteristic pairs to find all possible solutions. Prompted by the IKMDB, the CSR asks the narrowing questions, varying the order of entering processes, relationships, etc.

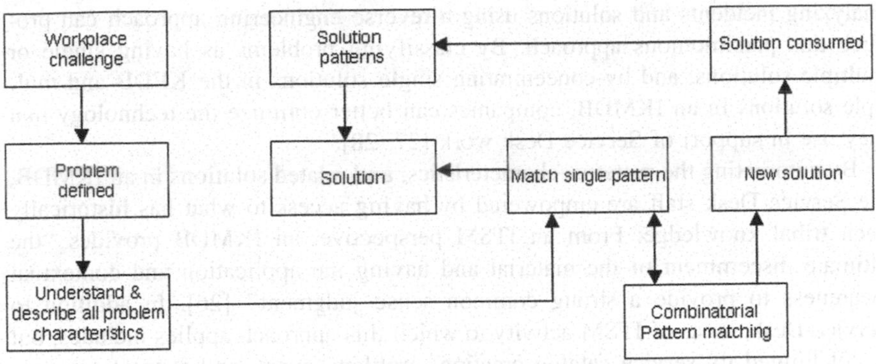

Fig. 4 Incident knowledge management database (IKMDB) in use

based on the answer to the prior question. As questions are answered, the data is entered into the IKMDB for reasoning on which question type to ask next and also to narrow the possible solutions. If the solution set contains a single answer, questioning stops and the solution is given. The IKMDB continues matching incident characteristics based on answers to the questions to eventually arrive at the incident definition with a correct solution set. Because of the nature of IKMDB problems, that is, having more than one solution to a specific issue, the solution set might contain multiple possible correct solutions that would then be reviewed to determine which solves the issue.

If no solution-incident characteristic set matches are found, the CSR troubleshoots and maintains a log of the actions that resulted in his eventual success. Upon CSR testing to identify a successful solution, a new solution-incident characteristic set is identified. As part of the resolution activity, the CSR enters pertinent information into the IKMDB and a new set of elements is created.

A very complex incident is one for which sub-incident components match patterns from several prior cases. In this case, recursive pattern matching might be conducted to identify all incidents, elements, contexts, or characteristics that match this particular issue. This situation requires novel combinatorial pattern matching to reach a workable solution. Thus, upon confirmation of successful resolution many IKMDB records would be updated with new components, characteristics, and contexts to document the new combination of solutions that match this new incident.

4 Discussion and Limitations

It is desirable to automate as many incidents and their solutions as possible so that individuals in problem-solving situations can access past recurring patterns to produce optimal results with a minimum of time and energy. The approach to analyzing incidents and solutions using a reverse engineering approach can produce this parsimonious approach. By classifying problems as having single or multiple solutions, and by concentrating single solutions in the KEDB and multiple solutions in an IKMDB, companies can better optimize the technology that they use in support of Service Desk work [27, 28].

By automating the patterns, characteristics, and related solutions in an IKMDB, the Service Desk staff are empowered by having access to what has historically been tribal knowledge. From an ITSM perspective, an IKMDB provides "the ultimate discernment of the material and having the application and contextual awareness to provide a strong common sense judgment" [26]. In addition to Service Desk, areas of ITSM activity to which this approach applies includes, but is not limited to, service catalog creation, problem, event, and request management, change management, release testing management, capacity planning, continuous service improvement decision making, continuity analysis, and others. In essence, any activity that requires complex problem solving is a potential candidate for IKMDB creation. Further, all problems encountered by a Service Desk, might ultimately find solutions in either a KEDB or an IKMDB, resulting in near 100 % FCR.

Many outcomes can occur from implementing a IKMDB to better support knowledge management of a Service Desk. Among possible improvement are increased user-perceived quality of service, reduced number of incidents requiring escalation, increased number of first call resolutions, and reduced number of Service Desk staff. The overall result of IKMDB implementation is satisfied customers, managers, and staff, who feel empowered and equipped with proper tools to excel at their jobs.

A database such as an IKMDB could be applicable to many types of knowledge work, Part of what makes knowledge work challenging is that it requires flexibility of reasoning, information from many sources, and a filtering of the information that eventually leads to an innovative solution. This type of reasoning seems to match that of an IKMDB closely. Thus, the reasoning processes described here that analyze situations from multiple perspectives, use the answers to questions to narrow the potential solution set, and develop solution sets applying Pareto analysis may solve many of the types of problems Davenport and McDermott described as being unautomatable at present [6, 27].

The obvious limitation is that what works in one location may not work in others. The Service Desk could be an isolated opportunity for applying KM to ITSM. We believe this is not an issue because the problems of pattern recognition and use of reasoning based on different types of problem characteristics occur in many ITSM situations as described above. In addition, the example of Service Desk management represents a general knowledge management problem that recurs across organizations. As a result, we believe the method of applying the DIKW model to develop an IMKDB is likely to be applicable in many settings, not just those relating to IT or ITSM.

A second limitation is the lack of empirical verification of the example in this paper. The expected improvements were developed from a case organization that required anonymity. The concept of pattern recognition that transcends different contexts is new to this work and, therefore, should be further verified through other case studies or through some type of empirical verification that compares, in this case, Service Desks, that differ in their methods of solution definition and finding. These are areas for future research.

Research on the concept of applying DIKW to knowledge workers is not new. The method of reverse engineering to develop the automated means of that application is new. Therefore, future research should focus on development of a generally applicable reverse-engineering process to enhance applicability beyond IT work.

Two aspects of this research for future research are the reverse engineering approach to reasoning from problem solution characteristics and the design and use of an IMKDB. While the concept of reverse engineering is not new, its application to solution-characteristic sets is new. Further, the actual steps of reverse engineering are not well articulated, nor are contingencies or alternatives known beyond practitioner tribal knowledge. Therefore, these are both areas for future research.

Similarly, the concept of an IKMDB is novel to this research. As a result, the scope, universality of applicability, methods of codification, storage, and presentation, best practices in logical and physical database design, application of intelligent search, and results descriptions all are subjects for further research. An approach to situational understanding, such as case-based reasoning, might further improve the success of IMKDB applicability [22]. Contingencies, alternatives, areas requiring expert reasoning and approaches to that reasoning for each research area are potential topics. Thus, there is a wealth of information to be discovered about how best to develop and use an IKMDB.

Eventually, another type of application may work prior to involvement of a Service Desk by monitoring applications in their operational environment for preventive maintenance. This preventive maintenance capability would identify issues with operation environments before they become incidents. Smart software daemons in the operations and Service Desk environments might be used to identify trends and situational differences that are potential problem causes.

Another type of research might approach the design of Service Desk automation support from a design science [17] point of view. In addition to dealing with multinational and cultural issues, design science ensures that value accrues from the activity and that it is appropriate to its environment. Thus, a design science approach, might result in further automatability of knowledge management activities.

5 Conclusions

Building and using an IKMDB can save significant numbers of staff, can speed many types of situational analysis that have relied on experience and tribal knowledge in the past, and can improve quality of service through these activities.

As a result, KM is critical to obtaining cost savings that are the promise of ITSM improvements. An IKMDB provides the ability to apply prior experience to solution of a new problem by automating information objects that provide clues to how new problems might use combinations of elements from prior solutions. The IKMDB thus, provides knowledge-based support that exceeds to capabilities of current KEDB information support. An IKMDB appears capable of supporting the development of skills in Service Desk staff in addition to improving the staff's ability to solve an ever-broadening range of incidents and the challenges in resolving them efficiently. Finally, the IMKDB discussed in this research offers an example to other types of knowledge work and how a reverse engineering approach might increase returns from the present method of analysis.

By classifying incident-solution sets into one of the four types, both KEDBs and IKMDBs can be customized to concentrate on the type of issue they best serve. By implementing an IKMDB to supplement a KEDB, companies can reduce staff and expenses, increase first call resolutions, and decrease overall time to reach a successful resolution to prior incidents. Thus, an IKMDB activity should be recommended for any relatively mature Service Desk organization and, by applying the concepts of the IKMDB to other types of knowledge work, automated support might be possible.

References

1. Ackoff, R.L.: From data to wisdom. J. Appl. Syst. Anal. **16**(1989), 3–9 (1989)
2. Adler, M.J.: A Guidebook to Learning:for a Lifelong Pursuit of Wisdom. Macmillan, New York (1986)
3. Alavi, M., Leidner, D.E.: Review: knowledge management and knowledge management systems: conceptual foundations and research issues. MIS Q. **25**(1), 107–136 (2001)
4. Bardhan, I., Demirkan, H., Kannan, P., Kauffman, R., Sougstad, R.: An interdisciplinary perspective on IT services management and service science. J. Manag. Inf. Syst. **26**(4), 13–64 (2010)
5. Conger, S.: Introduction to the special issue on IT service management. Inf. Syst. Manag. **27**(2), 100–102 (2010)
6. Davenport, T.H.: Rethinking knowledge work: a strategic approach. McKinsey Quarterly.www.mckinseyquarterly.com/ Rethinking_knowledge_work_A_strategic_approach_2739 (2011). Accessed 3 Feb 2011
7. Davenport, T. H., Prusak, L.: Working Knowledge. Harvard BusinessSchool Press, Boston (1998)
8. DeBono, E.: Thinking Course: Powerful tools to transform your thinking. BBC Active, London (1996)
9. Drucker, P.F.: The Essential Drucker: The Best of 60 Years of Peter Drucker's Essential Writings on Management. Harper Paperbacks, NY (2008)
10. Fricke, M.: The knowledge pyramid: a critique of the DIKW hierarchy. J. Inf. Sci. **35**(2), 131–144 (2008)
11. Gan, Y., Zhu, Z.: A learning framework for knowledge building and collective wisdom advancement in virtual learning communities. Edu. Technol. Soc. **10**(1), 206–226 (2007)
12. George, M., Kay, D., Oxton, G., Thorp, D.: Knowledge Centered Support (KCS) v5 Practices Guide. Consortium for Service Innovation, San Carlos (2010)

13. Gold, A.H., Malhotra, A., Segars, A.H.: Knowledge management: An organizational capabilities perspective. J. Manage. Inf. Syst. **18**(1), 185–214 (2001)
14. Goldberg, E.: The Wisdom Paradox: How Your Brain Can Grow Stronger As You Grow Older. Gotham, NY (2006)
15. Hahn, K.: Section 11: Methods of integration. http://www.karlscalculus.org/calc11_5.html (2010).Accessed 22 Nov 2010
16. Hawkins, J.: On Intelligence. Owl Books, NY (2005)
17. Hevner, A.R., March, S.T., Park, J., Ram, S.: Design science in information systems research. MIS Q. **28**(1), 75–105 (2004)
18. Jennex, M. (ed.): Current Issues in Knowledge Management. IGI Global, Hershey, PA (2008)
19. Kakabadse, N.K., Kakabadse, A., Kouzmin, A.: Reviewing the knowledge management literature: towards a taxonomy. J. Knowl. Manag. **7**(4), 75–91 (2003)
20. Kelley, G.: Selected Readings on Information Technology Management: Contemporary Issues. IGI Global, Hershey (2009)
21. King, W.R., Marks Jr, P.V.: Motivating knowledge sharing through a knowledge management system. Omega **36**, 131–146 (2006)
22. Kolodner, J.: Case-Based Reasoning. Morgan Kaufmann Publishers, San Francisco (1993)
23. Lee, H., Choi, B.: Knowledge management enablers, processes, and organizational performance: an integrative view and empirical examination. J. Manag. Inf. Syst. **20**(1), 179–229 (2003)
24. Leonard, D., Sensiper, S.: The role of tacit knowledge in group innovation. California Manag. Rev. **40**(3), 112–132 (1998)
25. Machlup, F.: Semantic quirks in studies of information. In: Machlup, F. Mansfield, U. (eds.) The Study of Information: Inter-disciplinary Message , Wiley, New York (1983)
26. Malone, T., Menken, I., Blokdijk, G.: ITIL V3 Service Capability RCV - Release, Control and Validation Best Practices Study and Implementation Guide. Emereo Publishing, Brisbane (2008)
27. McDermott, R.: Why information technology inspired but cannot deliver knowledge management. California Manag. Rev. **41**(4), 104–119 (1999)
28. Nonaka, I., Byosiere, P., Borucki, C.C., Konno, N.: Organizational knowledge creation theory—a first comprehensive test. Int. Bus. Rev. **3**(4), 337–351 (1994)
29. Nanaka, I., von Krogh, G.: Tacit knowledge and knowledge conversion: controversy and advancement in organizational knowledge creation theory. Organ. Sci. **20**(3), 635–652 (2009)
30. Parasuraman, A., Berry, L., Zeithaml, V.: SERVQUAL: a multiple-Item scale For measuring consumer perceptions of service quality. J. Retail. **64**(1), 12–40 (1988)
31. Probst, J.: Knowing me, knowing you…ah ha: a knowledge management primer. In: Proceedings of Pink Elephant Annual Conference, Las Vegas, NV, 21–23 Feb 2010
32. Ritchie, G.: Incident Management–Do's and Don'ts. Livingston, Scotland: Seriosoft, Ltd. http://www.seriosoft.com (2011). Accessed 28 April 2011
33. Robertson, M., Scarbrough, H., Swan, J.: Knowledge creation in professional service firms: institutional effects. Organ. Stud. **24**(6), 831–857 (2003)
34. Schulze, A., Hoegl, M.: Organizational knowledge creation and the generation of new product ideas: a behavioral approach. Res. Policy **37**, 1742–1750 (2008)
35. Smith, K.G., Collins, C.J., Clark, K.D.: Existing knowledge, knowledge creation capability, and the rate of new product introduction in high-technology firms. Acad Manag J **48**(2), 346–357 (2005)
36. Soloway, E.: Learning to program = learning to construct mechanisms and explanations. Commun. ACM **29**(9), 850–858 (1986)
37. Tourniaire, F., Kay, D.: Collective Wisdom: Transforming Support with Knowledge. Colorado Springs, CO (2006)
38. Van Bon, J.: Foundations of IT Service Management Based on ITIL® V3. Van Haren Publishing, Netherlands (2008)
39. Van Grich, J.P., Churchman, C.W.: Applied General Systems Theory. Harper and Row Publishing, NY (1974)

40. Zack, M.: What Knowledge-Problems Can Information Technology Help to Solve. In: Hoadley E., Benbasat I. (eds.) Proceedings of the Fourth Americas Conference on Information Systems, 644-646, Baltimore, MD, August 1998
41. Zeithaml, V.A., Parasuraman, A., Malhotra, A.: A Conceptual Framework for Understanding e-Service Quality: Implications for Future Research and Managerial Practice, Marketing Science Institute, Issue 115 (2000)
42. Zins, C.: Conceptual approaches for defining data, information, and knowledge. J. Am. Soc. Inform. Sci. Technol. 5(4), 479–493 (2007)

Chapter 2
IT Governance in a Malaysian Public Institute of Higher Learning and Intelligent Decision Making Support System Solution

Abdul Rahman Ahlan, Yusri Arshad and Binyamin A. Ajayi

Abstract IT governance in public universities is yet to be fully appreciated. This is because IT governance is just evolving as a new paradigm to take care of the stakeholders' expectations in leveraging IT with business goals. This study explores issues of IT governance and the use of IDMSS in a public university in Malaysia. The interviews and documents were analysed to generate themes. The findings show the need for effective IT governance in the university. This will enable optimal usage of IT resources, which are expected to have an impact on the university's performance. Moreover, COBIT framework is presented to provide the benchmark of international standards and practices that the university can adopt in the future. Finally, two IDMSS methods are presented for application in this study. This empirical study on IT governance in the context of a public Institute of Higher learning in Malaysia attempts to document the role of IT and the evolution of its practices in the context of a university.

Keywords IT governance · COBIT · IT service management · IDMSS

1 Introduction

Information Technology (IT) is a term used to imply the infrastructure as well as the capabilities of organisations that establish and support it [10]. Nowadays, organisations have recognised their dependence on this infrastructure in their bid

A. R. Ahlan (✉) · B. A. Ajayi
Department of Information Systems, International Islamic University Malaysia, Jalan Gombak, 53100 Kuala Lumpur, Malaysia
e-mail: arahman@iium.edu.my

Y. Arshad
Universiti Teknikal Malaysia Melaka, Durian Tunggal, Melaka, Malaysia
e-mail: ayusri@utem.edu.my

M. Mora et al. (eds.), *Engineering and Management of IT-based Service Systems*, Intelligent Systems Reference Library 55, DOI: 10.1007/978-3-642-39928-2_2,

to achieve their stated objectives, mission and goals and strategic positioning necessary for successful leverage. However, in the face of the reality expressed on the need to spend less on IT and to see IT rather as a utility [3], the animosity that greeted investments in IT became a thing of concern. In redressing this conception, lot of reactions came up and of such is the strategic implication of IT [2].

From this concept of strategic value, it is understood that IT has the potential to deliver strategic value at all levels, and this is not only restricted to the back office applications alone but rather to the innumerable ways in which IT is being used [2]. In fact, it is now more of a strategic partner [22] and what has been the concern today is the ability to utilise IT facilities to place organisations in their right perspectives. This ability has given birth to the concept of the management or better still the governance of IT facilities in order to allay the fear of stakeholders on IT investment given the spate of IT project failures. Thus, IT Governance (ITG) which has to do with structures, processes and relational mechanisms for the IT decision making in an organisation [20] should necessarily exist in any organisations that deal with IT, on a daily basis, though its quality (of existence) may vary from one organisation to the other [17].

Interestingly, educational institutions, which have been adjudged on the same footings as other organisations, are also concerned over the need to govern its IT facilities and resources in the business of service provision to 'customers' in this world that tends to move towards service orientation by the day. Researchers [21] have adjudged universities as having features typical of established organisations. This is a fact in today's business world whereby the service orientation paradigm has to do with the satisfaction directed from the organisation to the customers, the employees as well as all the stakeholders.

Malaysia is a nation in the South East Asia continent which has maintained a stable and growing gross domestic product (GDP) in the last 30 years. Its government is concerned about efforts geared towards achieving the status of a developed nation by the year 2020 and is not leaving any stone unturned in this regards. It is therefore, not surprising to see scheme like Malaysia Super-Corridor (MSC) rolled out to ensure compliance to standard [23]. The Institutes of Higher Learning are also encouraged to play vital role in this regards. That is why it is not surprising to see the case organisation used in this study joining the race of becoming a world-class university. Yet, ITG in it, is new, experimenting and in its infancy stage.

The purpose here is to present empirical findings of how ITG is being implemented in the university. The findings depicts: a university with unique vision and mission, striving to be a world-class university, still lacking in strategic emphasis on IT as an enabler. It's IT resources are decentralised and remain uncoordinated though this shortcoming has recently been identified and there have been concerted efforts to improve. Thus, being a 'service provider' in the Education industry, the university like its other counterparts has found IT handy to the achievement of its set objectives. The two main categories of information systems (IS) implemented in most Institutes of Higher Learning (IHLs), centre on:

1. Educational Learning Systems that now extends to online systems of instruction whereby real-time teacher-students interaction takes place. Today, there has evolved numerous ways of establishing the sharing of information within the university community members, inter-educational as well as the larger public.
2. Administrative Systems of running an organisation are now being deployed as never before, not-withstanding the bureaucratic nature of the university. Varied Information Systems has been deployed in the administration of the IHL; some are usually the one used in the industry and are adapted to use or those that are specifically developed to aid administrative efficiency in an academic setting. For example, enterprise integrations of resource's implementation of Balanced Scorecard are now gaining acceptance in IHLs.

The chapter is structured in the following orders. Section 1 introduces to readers on the main topic. Then, Sect. 2 provides a background literature on ITG concept, framework and its application in IHLs in Malaysian. Section 3 briefly describes the qualitative methodology adopted for the study while Sect. 4 provides detailed findings based on the analysis done. This is followed by Sect. 5 showing examples of how COBIT can be applied in the university environment in order to enhance its IT governance. Finally, the chapter ends with a conclusion, and future research is identified.

2 IT Governance in Institutes of Higher Learning

IT usage, from the foregoing realities, is becoming pervasive in this ever dynamic business environment and thus, its management and control cannot be left to chances. Such needs for the management of IT resources to afford leverage for businesses are necessary considering the huge investment that it now involves. This gives birth to the concept of IT Management, which focused on the effective and efficient internal supply of IT services and products and the management of present IT operations. This also takes into consideration the need for control and by extension, making of policies and rules of governance. It has made how IT resources are governed an important issue in the university community. As stated by Yanosky and Caruso [27] on the need for ITG in IHLs that there has been increasing attention to how IT is governed in the institutions over the past few years. One reason for these is the significant impact IT systems have on how institutions' works get done. In addition, a revival of interest in corporate governance, fed by financial scandal and a new wave of corporate accountability has also put a spotlight on how organisations of all kinds ensure that the expensive, complex, indispensable and strategy-enabling domain of IT is appropriately governed.

The popularity of ITG given this exposition has grown over the year [12] as IT has become the backbone of many businesses and organisations these days, relying more on its central role in the enterprise. Organisations with high levels of ITG could achieve more than 20 % greater profits than organisation with low implementation of the practices [25] and IHLs are no exception to this fact.

2.1 IT Governance Concept

The need for ITG, though a new concept [20], arose based on the fact that organisations have to think beyond IT and its infrastructures as a department, and that, there should be justifications for budget committed to the sustenance of such IT products and services. Thus, given this scenario, ITG, concentrates on performing and transforming IT to meet present and future demands of the business process (internal focus) and business customers (external focus). There have evolved various ways of defining ITG based on the perspective that is being considered, though, all revolve around the need to give direction and control on the use of IT and its resources. A review of some definitions prominently available in the literatures is provided in Table 1.

In summary, ITG is the apprehension experienced when exercising the rights on decision or policy-making: a function embedded in the corporate governance of an organisation, and the plan coupled with the will to carry out such decision on the structures and processes in executing organisations' goals. For this study, we shall use an operational definition adopted from these definitions ITG could be seen as: The control exercised by the Board of Directors and Managements in the formulation of right decisions, which enables the establishment of proper structures and processes thereby upholding best practices in order to enable strategic alignment of IT with the business thus ensuring that the organisations' IT sustains and extends its strategies and objectives in aiming for optimum usage well felt by its managed users which leads to good performance.

2.2 IT Governance Framework

A model that is usually referred here [20] is accompanied by a list of 33 best practices delineated as; 12 structures, 11 Processes, and 10 Relational Mechanism. In each of the layers of the model (strategic, management and operational), there are structures, processes and relational mechanisms (see Fig. 1).

Structure include organisational units and roles responsible for making IT decisions and for enabling contacts between business and IT management's (decision-making) functions (e.g. steering committees) [19]. This is best appreciated as an outline of how the Enterprise Governance of IT will be structurally organised. For example, does the CIO relates directly with the Executive Board or has to report to a CEO, who, in turn re-present the CIO's view at the Board?

Process means to formalise, and institutionalise IT strategic decision-making or IT monitoring procedures. Accordingly, this is to ensure that the day-to-day

Table 1 IT governance definitions

Author(s)	Definitions	Import(s) from definitions
ITGI [10]	"…The responsibility of the board of directors and executive management. It is an integral part of enterprise governance and consists of the leadership and organisational structures and processes that ensure that the organisation's IT sustains and extends the organisation's strategies and objectives"	Alignment of IT with the enterprise and realisation of the promised benefits Use of IT to enable the enterprise by exploiting opportunities and maximising benefits Responsible use of IT resources Appropriate management of IT-related risks
Weill and Ross [26]	Specifying the decision rights and accountability framework to encourage desirable behaviour in using IT	Cost-effective use of IT Effective use of IT for asset utilisation Effective use of IT for growth Effective use of IT for business flexibility
Van Grembergen et al. [20]	ITG is the organizational capacity exercised by the Board, executive management and IT management to control the formulation and implementation of IT strategy and in this, way ensure the fusion of business and IT	Enterprise Governance of IT: the sound management of IT as part of Corporate Governance, the responsibility of the board of directors, business management, and IT management to a merger between business and IT to come Strategic Alignment: aligning business strategy, IT strategy, business processes and IT processes so that the objectives of the IT organisation support optimal Value Creation: realising business value from IT-enabled business investments
Webb et al. [24]	The strategic alignment of IT with the business such that maximum business value is achieved through the development and maintenance of effective IT control and accountability, performance management and risk management	ITG is seen as evolving to be part of the Corporate Governance (the ways companies are directed and managed in a bid to monitor risk, meet set objectives and achieve an optimised performance) especially when the concept of Strategic Information System Planning (utilisation of information technology as a source of competitive advantage, and as a means of enabling and directing strategic moves of an organisation) is adhered to [6]

dispositions are consistent with the policies and provides input back to decisions for example [19]; the IT Balanced Scorecard (BSC was a management way of measuring organisation performance as popularised by Kaplan and Norton [11].

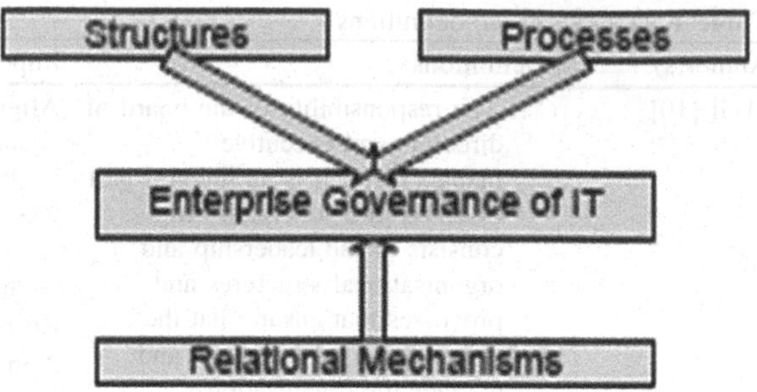

Fig. 1 The three layers of enterprise IT governance [20]

This involves supporting the business to be executed in an orderly manner; e.g. what are the interdependence of key performance indicators and the reporting and evaluation processes? Relational mechanism means active participation and collaborative relationship between Corporate Executives, IT as well as Business Managements. This includes announcements, advocacies, channels and education efforts [19].

2.3 IT Governance in Malaysian Higher Learning Institutions

It is an acclaimed fact that information plays a major role in the delivery and creation of knowledge (McRobbies and Palmer as cited in [9]) especially in knowledge-based environments like a university. However, empirical study on the influence of ITG on IT success in public organisations is scarce and worse still when the focus is on the HEIs. This is likely due to the recentness of the phenomenon and the lack of clear understanding of the concept of ITG [14]. An example is a recent study in a Malaysian university [9] where he observes that despite the IT infrastructures on ground, the university, in question, lack proper IT planning team structure. This goes to show that despite realising the pervasiveness of the IT yet little credence is given to its proper governance.

3 IT Governance in Institutes of Higher Learning

We adopted a qualitative study fashioned along on a single case organisation. The technique of data collection was through interview and extractions of some facts from the archives of the organisation. This was done with the mind of finding out: what is "in and on someone else's mind" or to find out from them those things we cannot directly observe… We have to ask people about those things that cannot be ordinarily observed in order to allow us to enter into their perspectives [15].

Interview sessions were held with the Director of IT function that oversees all affairs of IT in the university, and with Head of ITG department with questions prepared beforehand which, "should not exceed ten…and has the advantage of allowing a certain degree of flexibility and allows for the pursuit of unexpected lines of enquiry during the interview" [8]. These questions were both structured in a way to gain insight to the followings:

1. The impact of the IT structure on ITG practices in the university.
2. The level of managements' control on IT resources in the university.
3. The level of impact of IT structure and management control on ITG.

4 Research Findings

4.1 Case Background

It is an acclaimed fact that information plays a major role in the delivery of services. The university was established in the 1980s with the vision of becoming a leading international centre of educational excellence that seeks to restore the dynamic role of its products in all branches of knowledge. It started at a mini campus but now has a main campus where central administrative activities take place with a host of academic programmes. The university, aside from retaining the maiden campus also resides in other campuses.

Through generous support from Malaysian Government, its facilities are kept up-to-date to the changing demands of its core businesses; teaching and learning, research and consultancy. It operates under the directive of a Board of Governors with representatives from government and other international bodies. As of December, 2012, it has the growing population of over 1,900 academic staff and more than 1,800 non-Academic staff with Student's population of over 22,000 undergraduate and more than 5,000 postgraduate.

Given the spread of the campuses and the attendant use of IT resources to impact knowledge and affect the lives of the citizenry by community services, thus the need for ITG cannot be overemphasised. This is because all the university's facilities and establishments run on IT-enabled systems to afford better management. The need for governance is enhanced by the need to use resources of IT to enable this educational essence of the university.

4.2 Themes' Generation

The themes that were uncovered from the interviews, learning from the wealth of knowledge and experiences of the respondents in the field, are best appreciated within the context of data analysis of a case study as enumerated thus:

1. Organisation of detail about the case.
2. Categorisation of data.
3. Interpretation of single instances.
4. Identification of patterns.
5. Synthesis and generalisations [5].

The data collected are interpreted and scrutinized digging out their underlying themes. These themes are carefully likened to the coding of the qualitative data; a process described… as a summative, salient, essence-capturing… [16] of data collected from interviews, participant observation field notes, artefacts, etc. and also such coding could be likened to analysis of the data [13].

The facts that were thus recorded were, firstly, organised orderly to reflect the objective of the study. The data were then categorised in such a way that they might make meaning and its specific information were analysed in each of the sections to throw light on such an instance. A sense of what the data meant was made by identifying the patterns at which they occur in the whole exercise and that informed their being grouped together to reflect that. Finally, data were mixed, separated and simplified to form a general idea, of how they apply to the case, exposure to facts at the university as well as what has been discovered through the exploration on the university's website as presented in Table 2.

No doubt change is a very difficult phenomenon to get by, but experience has shown that changing for better always leads to the desired goals and objectives. If only there would be a change on the management level to the way IT is been viewed as a department to its being seen as a partner in the business of the university, then to achieve and sustain the goal of university for world recognition would be less stressful. For that to be the case, all hands will have to be on deck to ensure that IT is seen as a partner in progress. From this study, we learn the following facts:

1. The management is yet to come to terms with the need to exercise control on the IT resources as things are left to the discretion of IT department.
2. This inability to appreciate the need for control has weighed more on the IT department as it is doing much but has less impact.
3. The decision making on IT resource's usage is at the mercy of individual champions in their respective faculties/functions because of the decentralised IT structure. Thus making it difficult to measure optimality of their usage at the university level to determine how objectives of the university are being affected by commitment on IT.
4. There is need for authority to support a university-level IT policy with its implementation on an enterprise-wide scale.
5. The IT function needs to achieve buy-ins of stakeholders as well as the required authority from management for IT policy to be effective in the university. This will ensure that users carry the IT function along in their IT investments as it will be prioritised in line with the university business processes, and all systems will be harmonised together.

Table 2 Summarised issues emanating from the transcribed interviews and archive

Theme	Issues	Remarks
Role and structure	The role of IT	Relegated to advisory role or order takers
	Impact of structure on ITG	No central coordination of IT resources
	Empowerment and awareness	No distinct ownership of processes
		Role of CIO not well felt
		Effective efforts from management on creation of awareness for adoption of technology is not seen
		The IT function seeks the buy-ins of stakeholders so that best practices can be entrenched
		Need for awareness on understanding business process before going for IT facilities
Top management perspectives	Management presence in IT committees, and decision making	Bulk of responsibilities pass on to the IT
	Role and impact of the CIO	There is lack of platforms to discuss IT problems
		Biasness to IT issues
		The need for the office of CIO and the filling of the post over time shows a need to "fill in the gap" in order to abide with the government's directives
		There has been a communication breakdown between CIO and reality on ground at the IT level
Challenges in enabling IT resources' governance	Effective usage of resources	Decentralised IT structure and budget: IT planning and procurements not necessarily dictated by business process needs
	IT policy formulation control and enforcement	Effective measure of IT asset usage has been hindered
	The use of tools and best practices	The users sees the IT policy as IT's department policy
	Users' relations: Feedback mechanisms	Difficulties in measuring the feedback is because of the ad-hoc nature in the way things are being presently done
		Dearth of manpower to handle and propose better methods to the situation on ground
	Manpower management	Need for coalition among all IT staff of the university

6. There is a need to ensure that accountabilities on IT resources are put into university functions, in their interaction with IT. This will evolve a university-level adoption of benchmarking control's frameworks like ITIL, ISMS, COBIT, Val IT, BSC IT.
7. There is need for top management to define the road map whereby processes are assigned their respective owners.

5 Aligning IT Governance to COBIT Framework

Control Objectives for Information and related Technology (COBIT) [4] and ISO 17799 have been used as reference frameworks for Information Security governance and hence, ITG in real practices. This section, however, will limit and only delve on COBIT as a reference framework while at the same time acknowledging the importance of other international reference standards such as ISO 17799.

COBIT is a 'tool for information technology governance' [4]. COBIT is therefore, not exclusive to information security—it addresses Information Technology governance, and refers among many other issues, to information security. COBIT divides Information Technology governance into 34 processes, and provides a high level Control Objective (CO) for each of these 34 processes. Each CO is again divided into a set of Detailed Control Objectives (DCOs), which specify the way the high level CO must be managed, in more detail. In total, 316 DCOs are defined for the 34 processes. The rationale is that if each of these 34 processes is managed properly, good Information Technology governance will result. Nevertheless, COBIT is in many cases preferred by IT auditors and IT Risk Managers as a framework of choice.

The COBIT framework contributes to these needs by:

- Making a link between the business requirements and IT through goals,
- Defining the responsibilities for obtaining the goals,
- Organizing IT activities into a generally accepted process model,
- Identifying the major IT resources to be leveraged,
- Measuring the maturity levels of IT processes,
- Defining the management control objectives to be considered.

COBIT defines IT activities in a generic process model within four domains. These domains are Plan and Organise, Acquire and Implement, Deliver and Support, and Monitor and Evaluate. The domains map to IT's traditional responsibility areas of plan, build, run and monitor. The COBIT framework provides a reference process model and common language in an enterprise to view and manage IT activities. Incorporating an operational model and a common language in all parts of the business involved in IT is one of the most important and initial steps toward good governance. It also provides a framework for measuring and monitoring IT performance, communicating with service providers and integrating best management practices.

A process model encourages process ownership, enabling responsibilities and accountability to be defined. To realise the IT strategy, IT solutions need to be identified, developed or acquired, as well as implemented and integrated into the business process. In addition, changes in and maintenances of existing systems are covered by this domain to make sure the solutions continue to meet business objectives.

The application of COBIT in an organisation varies on its IT systems and infrastructure's complexities. Depending on the formulation of an organisation's information system, an auditor assesses on all control objectives outlined in COBIT using any assessment methods on the IT general and application controls. For small and medium-sized organisations, it is unlikely that the conditions prevail in big organisations, which are more complex. Nevertheless, with the application of COBIT in strategic, tactical and operational levels, many organisations have benefited from their successful formulation and implementations.

6 Intelligent Decision-Making Support Systems for IT Governance

Control IT governance is a broad concept and a comprehensive framework is needed to account for it. With thousands of input nodes and decision trees that a system must calculate to find the best solution path, an artificial intelligence or fuzzy logic based on a mathematical formula or expert system has become a feasible option. Nonetheless, management must know how to operate the system as well. The ITG paradigm aims at providing the decision-making structures, processes, and relational mechanisms, needed in order that IT supports and perpetuates the business. Hence, formulating an intelligent decision making support system (I-DMSS) for ITG is essential in order to capture all important scopes for management decision making. Management only needs to see the core criteria embedded in the system and consequently, decide on the best solution. The adjacent discipline of enterprise architecture provides a broad range of frameworks and tools for model-based management of IT. Enterprise architecture is a commonly and successfully used approach, but the frameworks need to be adapted with respect to the concerns at stake in order to become truly useful [18].

Decision-making Support systems (DMSS) are Information Systems designed to interactively support all phases of a user's decision-making process. There are various notions about all aspects of this definition. There can be individual, group, and other users. Support can be direct or indirect. The decision-making process can be viewed in various ways. User-computer interaction can have a variety of dimensions. The information system offering the support can involve many technologies drawn from several disciplines, including accounting, cognitive science, computer science, economics, engineering, management science, and statistics, among others.

Many approaches for ITG decision-making support systems can be applied. We recommend here two methods or approaches that can be applied for designing, developing and constructing ITG IDMSS. One method is using mathematical foundation of the prediction apparatus that is using a Bayesian network which is based on statistical data to support ITG decision making. The method relies on best-practice frameworks, including the Control Objectives for Information and related Technology, COBIT and Weill and Ross [26] method for ITG performance assessments. In addition, the method takes ITG two steps further. Firstly, it combines ITG with a modelling approach inherited from the enterprise architecture discipline. Secondly, it provides information about statistical correlations between an IT organisation's internal effectiveness, and the effect thereof as perceived by business stakeholders. The method features an ability to assess ITG maturity and performance. Further, it incorporates a framework for prediction of ITG performance given ITG maturity, which makes it suitable for decision making support concerning IT organisation matters [18].

Another approach to designing, developing and constructing IDMSS systems is by using the logic-semiotic apparatus. Golovina [7] proposed an approach to designing the IDMSS based on Pospelov's theory of semiotic modelling in comparison of Bayesian network approach described earlier. The SETRIAN modelling subsystem is based on Trincon version 9.0. It comprises four main components, including:

1. Subsystem for modelling the controlled object;
2. Subsystem of strategic planning for defining the initial and final states of the controlled object; planning the ways to reach the target state of the controlled object; choice of the best way to reach the target state of the controlled object;
3. Infological subsystem used to define the ontology of the knowledge domain (objects, relations, axiomatic), as well as to define the structure of the controlled object, information about its modes of operation, and so on; and
4. Diagnostic subsystem explaining the causes of emergencies at the controlled object and monitoring the outcomes of the decisions made in the course of modelling the dynamic controlled object.

The developed fuzzy-model editor is distinguished by using methods for extraction of imprecise fuzzy information, neuron networks for extrapolation of values upon constructing the membership functions, modifiers for obtaining new membership functions (values of the linguistic variables), and parallel representation of the membership functions and support of the 3D-graphics. These functional capabilities enable one to construct in shorter time the membership functions that are adequate to the expert's judgment of the operation of the controlled object [7].

There are several versions of the Trincon development tool for fuzzy decision-making support systems which each is a functional extension of the older version. Developed in the late 1999, Trincon version 9.0 is an integrated tool for developing fuzzy decision making support systems. The Trincon version 9.0 comprises the:

1. Inference system based on the generalised Modus Ponens adjusted to the Tnorms and T-conorms;
2. Subsystem for loading the source data such as the linguistic variables whose values are described by the membership functions and fuzzy production rules, as well as the initial tests;
3. Subsystem for logging the operation of the fuzzy control system developed using Trincon; subsystem for adjusting inference to the hybrid T-norms;

4. Subsystem for visualising operation of the decision-making system and supporting 2D- and 3D-graphics; subsystem of semantic modification of the T-norm;
5. Subsystem supporting semiotic modelling based on fuzzy logic;
6. Information-and- control system providing the decision maker with missing information;
7. Multitasking-support subsystem; and
8. Help subsystem [7].

The most important component of the IDMSS is the information extraction and predictive capability. Information extraction is the capability to extract the useful information from the abundance of information. This information will serve as the input to IDMSS that will be presented to the user in a meaningful format. The IDMSS predictive capability will use and analyse the information into a pattern which represents a trend of the event. This trend will be learned and will be used to predict the future event. These facilities exhibit intelligent behaviour of the IDMSS and are very useful in assisting the decision maker.

Hence, the second approach provides an IDMSS for ITG that can be adopted by the organisation. The IDMSS can solve the ITG themes or areas in intelligent manners by using predictive capability for example. Senior management needs a system that can trigger the key points for them to make decisions. Using IDMSS will summarise important information for decision making for ITG. Each issues emanating from every single unit, centre or faculty will be handled by the fuzzy logic in Trincon, for instance. Then, a few triggers can be generated from the fuzzy IDMSS in order to assist the senior management in their ITG decision making.

7 Conclusion

The contribution of this study on ITG in Malaysian universities, its limitation and future works are presented.

Without doubt, universities combine some of the features of both private (profit-based) and public (non-profit) organisations' orientations in that though they are not entirely profit-making but have set targets to achieve while making good their contribution to humanity. It is also independent because of degree of freedom on its way of doing things, and this has a lot on its output. With this in mind, this study shows the way such a unique organisation will have to be dealt with. It is an attempt that represents an empirical study on ITG in the content of Institutes of Higher Learning Institution in Malaysia.

For a university is expected more in terms of practicing the best and making good use of knowledge. It is a case of how wisdom is a lost treasure of the Wise one, so he is its rightful owner at its discovery. Thus, this study portends that:

1. The management needs a renewed interest in the systematic way of exercising control on the IT resources so that the IT department will be able to govern the resources effectively with their support.
2. The IT department needs to appreciate that getting stakeholders' buy-ins for best practices in an educational outfit is important. The leadership has to understand that it is not a case of one cap fits all. It is one thing to think of new ways of doing things; it is another thing to get the buy-in from the stakeholders as its adoption will affect ways of doing things [1].
3. The university members need to appreciate that adopting the right course in dealings with IT resources will imply contributing one's quota to knowledge and self-satisfaction as members of the community and the nation at large.
4. More research and development on IDMSS needs to be done in public and private Institutes of Higher Learning in Malaysia.

There is still a room for improvement on the current study as it is limited by the fact that:

1. It is a single case study; so scientific generalisation may not be all that possible [28].
2. The perspective of the users is yet to be explored. The users constitute a good proportion of the community, and their input would help form a holistic view of the ITG in the organisation and a contribution to the body of knowledge.
3. We restrict ourselves to interview, available facts from the archives and the website as a form of technique in gaining insight to the raised questions.
4. The proposed conceptual IDMSS needs to be applied to the university ITG environment in order to test each technique's suitability and efficiency.

We suggest a further study that can complement the present one as a way of extending the search light to the users of the IT facilities. It is also hoped that an elaborate comparative study of a multi-case study research could be conducted further on Malaysian Universities to see how university's processes have become stable, to the extent that they can enjoy some further benefits over their contemporaries. Here, there is a need to share experiences. In addition, each IDMSS system needs to be tested in more sample sites so that the system developed will be more inclusive and robust for all environments.

References

1. Ajayi, B.A., Badi, B., Al-ani, M., Dahlan, A.: Taking community-based system to Malaysian communities for disaster management. Int. J. Human. Soc. Sci. **1**(7), 171–177 (2011)
2. Bannister, F., Remenyi, D.: Why IT continues to matter: reflections on the strategic value of IT. Electron. J. Inf. Syst. Eval. **8**(3), 159–168 (2005)
3. Carr, N.: IT doesn't matter. Harv. Bus. Rev. **81**, 41–49. (2003)
4. Control Objectives for Information and Related Technologies (COBIT), 3rd edn. IT Governance Institute (ITGI), Rolling Meadow, IL, USA (2000)

5. Creswell, J.W.: Educational Research: Planning, Conducting, and Evaluating Quantitative and Qualitative Research, 3rd edn. Merrill, Upper Saddle River (2008)
6. Galliers, R.D.: Strategic information systems planning: myths, reality and guidelines for successful implementation. Eur. J. Inf. Syst. **1**, 55–64 (1991)
7. Golovina Elena Y.: An approach to designing intelligent decision-making support systems on the basis of the logic-semiotic apparatus. In: Proceedings of the 2002 IEEE International Conference on Artificial Intelligence Systems (ICAIS'02) (2002)
8. Grix, J.: The Foundations of Research. Palgrave Macmillan, New York (2004)
9. Ismail, N.A.: Information technology governance, funding and structure: a case analysis of a public university in Malaysia. Campus-Wide Inf. Syst. **25**, 145–160. (2008)
10. ITGI.: Board Briefing on IT Governance. IT Governance Institute, Rolling Meadows (2003)
11. Kaplan, R.S., Norton, D.P.: The balanced scorecard-measure that drives performance. Harv. Bus. Rev. **70**, 71–79 (1992)
12. Marrone, M., Hoffman, L., Kolbe, L.M.: IT executives' perception of CobiT: satisfaction, business-IT alignment and benefits. In: Proceedings of the Sixteenth Americas Conference on Information Systems, pp. 1–10. Lima, Peru (2010)
13. Miles, M.B., Huberman, A.M.: Qualitative Data Analysis: An Expanded Sourcebook, 2nd edn. Sage, Thousand Oaks (1994)
14. Pardo, T.A., Gil-Garcia, R.J., Burke, B.G.: Governance structures in cross-boundary information sharing: lessons from state and local criminal justice initiatives. In: Proceedings of the 41st Hawaii International Conference on System Sciences, pp. 211–220. IEEE, Waikoloa Hawaii (2008)
15. Patton, M.Q.: Qualitative Evaluation Methods. Sage, Thousand Oaks (1990)
16. Saldana, J.: The Coding Manual for Qualitative Researchers. Sage Publication, London (2009)
17. Simonsson M., Johnson, P., Ekstedt, M.: The effect of IT governance maturity on IT governance performance. Inf. Syst. Manag. **27**, 10–24 (2010)
18. Simonsson, M.: Predicting IT Governance Performance: A Method for Model-Based Decision Making. Unpublished doctoral thesis. Industrial Information and Control Systems KTH, Royal Institute of Technology Stockholm, Sweden (2008)
19. Van Grembergen, W., De Haes, S.: Enterprise governance of IT in practice. In: Van Grembergen W. (ed.) Enterprise Governance of Information Technology, pp. 21–75. Springer, New York, USA (2009)
20. Van Grembergen, W., De Haes, S., Guldentops, E.: Structures, processes and relational mechanisms for IT governance. In: Van Grembergen, W. (ed.) Strategies for Information Technology Governance, pp. 1–36. Idea Group Publishing, Hershey (2004)
21. Vathanophas, V., Stuart, L.: Enterprise resource planning: technology acceptance in Thai University. Enter. Inf. Syst. **3**(2), 133–158 (2008)
22. Venkatraman, N.: Valuing the IS contributions to business. Comput. Sci. Corp (1999)
23. Vicziany, M., Puteh, M.: Vision 2020, the multimedia supercoridor and Malaysian universities. In Cribb, D.R. (ed.) Proceedings 15th Biennial Conference of Asian Studies Association of Australia, pp. 1–20. Canberra, Australia (2004)
24. Webb, P., Pollard, C., Ridley, G.: Attempting to define IT governance: wisdom or folly? In: Proceedings of the 39th Hawaii International Conference on System Sciences, pp. 1–10. IEEE , Hawaii (2006)
25. Weill, P., Ross J.: Designing IT governance. MIT Sloan Manag Rev. **46**(2), Winter (2005)
26. Weill, P., Ross, J.W.: IT Governance-How Top Performers Manage IT Decision Rights for Superior Results. Harvard Business School Press, Boston (2004)
27. Yanosky, R., Caruso, J.B.: Process and politics: IT governance in higher education. Educause Cent. Appl. Res. 1–10 (2008)
28. Yin, R.K.: Case Study Research: Design and Methods. Sage, Thousand Oaks (2003)

Chapter 3
DSS Based IT Service Support Process Reengineering Using ITIL: A Case Study

Raul Valverde and Malleswara Talla

Abstract The Information Technology Infrastructure Library (ITIL) is readily available for establishing the best practices, reengineering and improving the IT service support process. However, the ITIL framework only provides recommendations, and a company needs to explore a methodology for improving the IT service support process and adopting the best guidelines of ITIL framework. To this end, this chapter investigates upon how to apply the ITIL framework can be used for evaluating the current IT service support process and its reengineering. A set of Key Performance Indicators (KPI) were established which are monitored by a decision support system (DSS) for triggering on-going reengineering of IT service support process. A case study methodology is used for an effective reengineering of IT service support process. This chapter focuses on implementing the ITIL guidelines at an operational level, improving the service desk, incident management, problem management, change management, release management, and configuration. It also focuses on implementing the ITIL guidelines at a tactical level, improving the service level management, capacity management, IT service continuity management, service availability, and security management. The chapter describes a methodology and an experience in implementing process reengineering techniques following ITIL framework.

Keywords ITIL · KPI · IT service support processes · DSS · Reengineering

---

R. Valverde (✉) · M. Talla
Concordia University, Montreal, Canada
e-mail: rvalverde@jmsb.concordia.ca

M. Talla
e-mail: mrtalla@jmsb.concordia.ca

M. Mora et al. (eds.), *Engineering and Management of IT-based Service Systems*,
Intelligent Systems Reference Library 55, DOI: 10.1007/978-3-642-39928-2_3,

1 Introduction

The complexity of Information Technology (IT) applications makes it difficult to properly tune customer requirements and service provider capabilities. Customers often cannot express their real service requirements and do not know the corresponding performance needs. Likewise, service providers often do not know how to differentiate between IT services and how to attune them to a specific customer [1]. In order to address these problems, many organizations are adopting the Information Technology Infrastructure Library [2] as a framework to support their primary business processes. In the past, many IT organizations focused only on technical issues. Nowadays this has changed to a more service oriented way of thinking in order to provide a high quality service. IT became a part and participle of daily business activities in every organization.

The IT Infrastructure Library (ITIL) is not hardware or software, but it is a technique to manage the technology and communications in an optimal way. The ITIL is not a set of rules that must be followed, but a guideline to help organize and arrange the IT organization. The primary objective of the ITIL is to establish the best practices and improving the standard of IT service quality that customers should demand and providers should supply [2]. The ITIL can be used as a quality service guideline to help an organization to achieve the following objectives [2]:

- Better quality control,
- Increase service level,
- Cost reduction,
- Increase efficiency and effectiveness of information supply,
- Unambiguously describing the service in setting up Service Level Agreements (SLAs), and
- More control over business processes.

The ITIL plays an important role in helping a business organization to meet its objectives since it helps to manage the IT resources more efficiently. One should consider the environment (social, organizational, and physical), the processes and their interdependencies among different dimensions of an organization; e.g. clinical information system (CIS) and end-user of a bio-medical application, as depicted in Fig. 1 [3].

The ITIL framework for Service Delivery and Support can be accomplished in three levels: the strategic, tactical and operational level. The strategic level key performance areas aim at long term goals. The tactical and operational levels focus on medium and short terms respectively. Figure 2 presents the KPIs at a tactical and operational levels which are the main areas addressed in this research.

The basic premise of this work presented in this chapter is to investigate how to apply the ITIL framework for the reengineering the IT service support process. A set of key performance indicators (KPIs) for IT service support areas will provide a better means of monitoring need for reengineering. A decision support system (DSS) can gather data and derive the KPIs and monitor them in a timely manner

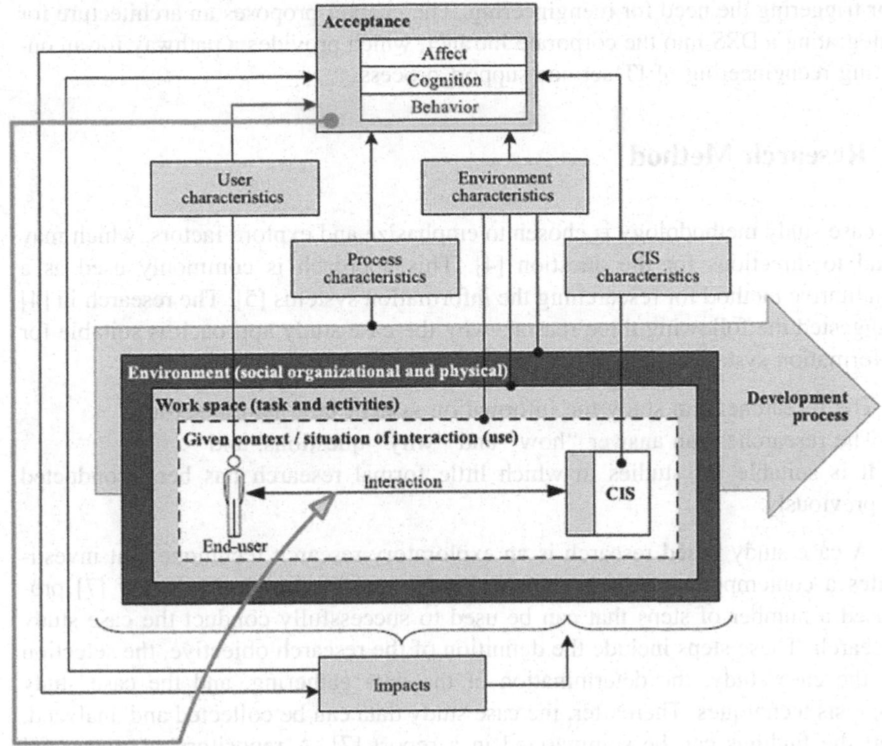

Fig. 1 The Information system interaction model (*Source* [3])

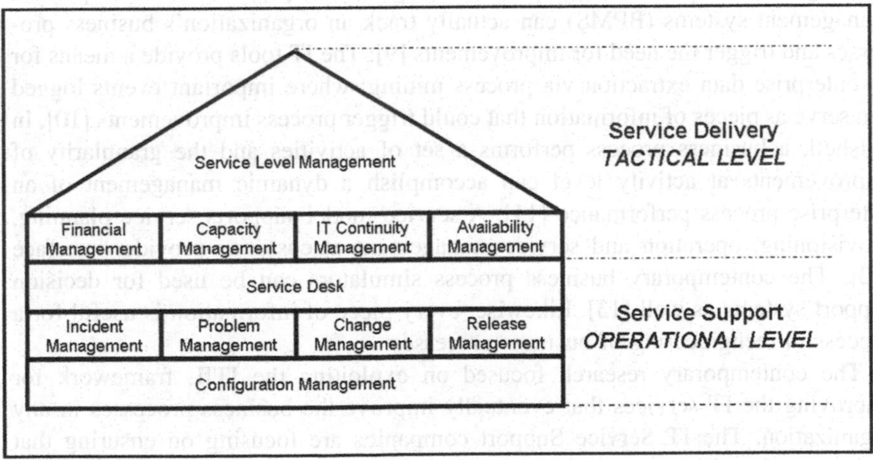

Fig. 2 ITIL core service management functions and processes (*Source* [30])

for triggering the need for reengineering. The chapter proposes an architecture for integrating a DSS into the corporate Intranet, which provides a pathway for an on-going reengineering of IT service support process.

2 Research Method

A case study methodology is chosen to emphasize and explore factors, which may lead to directions for the question [4]. This approach is commonly used as a qualitative method for researching the information systems [5]. The research in [4] suggested the following three reasons why the case study approach is suitable for information systems:

- The researcher can study the information system in a natural setting,
- The researcher can answer "how" and "why" questions, and
- It is suitable for studies in which little formal research has been conducted previously.

A case study based research is an exploratory research technique that investigates a contemporary phenomenon within its real-life context [6]. Soy [7] proposed a number of steps that can be used to successfully conduct the case study research. These steps include the definition of the research objective, the selection of the case study, the determination of the data gathering, and the case study analysis techniques. Thereafter, the case study data can be collected and analyzed, and the findings can be summarized in a report [7]. A repository of successful resolutions to the past problems can serve as a tool for preventing or solving the future problems, and each resolution can be visualized as a case while accomplishing an intelligent business process reengineering [8]. The business process management systems (BPMS) can actually track an organization's business processes and trigger the need for improvements [9]. The IT tools provide a means for an enterprise data extraction via process mining; where important events logged can serve as pieces of information that could trigger process improvements [10]. In nutshell, a business process performs a set of activities and the granularity of improvements at activity level can accomplish a dynamic management of an enterprise process performance [11]. A service model supports service planning, provisioning, operation and service management at customer-provider-interface [12]. The contemporary business process simulators can be used for decision support systems as well [13]. Likewise, every piece of information is useful for a successful reengineering of business processes.

The contemporary research focused on exploiting the ITIL framework for improving the IT services that eventually improve the business processes in any organization. The IT Service Support companies are focusing on ensuring that their customers have appropriate services to support for business functions [14]. The ITIL guidelines are widely used for improving IT service support processes [15]. Jantti [16] presents how the ITIL framework was used for improving the incident management processes in two companies. Just like all other business processes, implementing ITIL processes efficiently depends on building or procuring IT tools that can support them. The basic issue of supporting ITIL with process-oriented tools such as workflow management systems is presented in [17]. Implementing ITIL can be cumbersome and time consuming if one doesn't follow a roadmap for improvement based on priorities, dependencies, and guidelines [18]. The experience of IT staff and the amount of time devoted for understanding IT needs, and creating an appropriate service management office (SMO) could help improve the success rate of IT services [19]. A case study of managing IT services in finance industry by implementing suggestions that arise from ITIL methodology is presented in [20].

The objective of this research chapter is to investigate upon how to apply the ITIL framework for the reengineering of IT processes in an organization. The case study selected is of an IT services company located in Liverpool, UK. The selected company is currently providing several types of support services to many organizations in the UK. This research will concentrate in one of their customers, a company that is specializing in dental care. The material provided in this chapter with respect to the case study has the consent of the involved parties.

The research will start with a description of the case study, its organizational structure, main business services and client base. The data gathering is an important part in the case study research. In this regard, [21] and [6] identified seven sources of empirical evidence in case studies, as follows:

- Documents: Written material sources that include published and unpublished documents, company reports, memos, letters, agendas, administrative documents, departmental info, reports, e-mail messages, newspaper articles, or any document that presents some evidence of some information,
- Archival records: Archival documents can be service records, organizational records, and lists of names, survey data, and other such records,
- Interviews: An interview can be used for three purposes: as an exploratory device to help identify variables and relations; as the main instrument of the research; and as a supplement to other methods (Kerlinger 86). Interviews were conducted for the present study for the first and third purposes. As a method it is one of the most important sources of information for a case study: open-ended, focussed, and structured or survey. In this study various forms were combined for collecting the data.
- Questionnaires: These are structured questions written and supplied to a large number of respondents, commonly spread over a large geographical area for consideration in advance. Respondents fill in the blank spaces and return the questionnaires to the researcher either by post or in person. Sometimes inducements, such as a small gift, are used to encourage recipients to complete the questionnaires.
- Direct observation: This occurs when a field visit is conducted during the case study. This technique is useful for providing additional information about a topic being studied. Reliability is enhanced when more than one observer is involved in the task.

- Participant-observation: Participant-observation turns the researcher into an active participant in the events being studied.
- Physical artefacts: Physical artefacts can be tools, instruments, or some other physical evidence that may be collected during the study as part of the field visit. Use of a number of these instruments to obtain data from the same source provides for triangulation as defined in [22].

In this chapter, the case study uses the questionnaire, review of documents, archival records and observation techniques for collecting the data. The use of observation as a method of data collection is presented in [4, 21, 23] and it works well in a case research [6]. In this study, the researcher visited the site of information system to observe its functionality, collected several documents that identify the business processes and describe their current operations. This will help the researcher to learn about the details of the information systems included in the study.

Archival records are an integral part of the data that needs to be collected. The main records that will be used are the problems logs that are kept for future enhancements by the case study. These records will help the researcher to identify the areas of the IT services that will require modification for quality improvement. Based on the data collected, the researcher will perform a full analysis and benchmark the ITIL framework into the IT services operations. Further, a study on the effectiveness of ITIL framework will be conducted during the study case, in order to measure the improvement on the IT services after the ITIL framework implementation. To do this, a small portion of the ITIL framework will be implemented and one group pretest-posttest experiment will be conducted as suggested by [24].

The one group pretest-posttest experiment is a quasi-experiment in which the subjects in the experimental group are measured before and after the experiment is administered [24]. The participants of the experiment will be selected via convenience sampling. This sampling technique refers to obtaining sample units or people who are available [24]. This method is justified since the participation in the study will be voluntary and it is difficult to anticipate the number of participants in the sample. The key participants of the case study will be mailed an invitation letter.

Further on, a questionnaire will be used as a data-gathering device administered before and after the implementation of the ITIL framework in the case study. The questionnaire will be concise and effective in addressing the requirements of data, while considering the time and moneytary constraints. As a result, the questionnaire is defined as "a pre-formulated written set of questions to which respondents record their answers, usually within rather closely defined alternatives" [25]. Questionnaires have a number of inherent advantages in regard to conducting research. The most of significant of these are that they can be sent to a sample population that is dispersed over a wide geographical and they can be answered by respondents at their own convenience [26]. Furthermore, as the participants are assured complete anonymity, self-administered questionnaires overcome the problems of interviewer bias while reducing the respondent's likely reluctance to convey an incorrect or controversial information. Reliability is another advantage since the questionnaires are easily repeated [27]. That is, as respondents simply nominate a particular box to answer questions, no value judgments are required.

A simple-dichotonomy IT services evaluation questionnaire will be developed and administered to the participants: the pre-ITIL test (pretest) and post-ITIL test (posttest). A simple-dichotonomy question requires the respodent to choose two alternatives (Yes and No) [24]. Both tests will contain the same questions related to the IT services that need to be evaluated.

In order to analyze the data from the questions, and ascertain the general trends, descriptive statistics methods are used. The hypothesis that the ITIL framework helps to improve processes of the IT services is tested by using the *t* Test for comparing the mean values of the pre-ITIL test and post ITIL test [24] to find the evidence of a possible effect of the ITIL framework in improving the quality of services.

3 Process Analysis

The case study has ten dental clinics in different locations of Liverpool. All these clinics are connected via a high-speed Wide Area Network (WAN). The data is centralized into the IBM RS 6000 server located in the main dental center. Workstations are located in the user office and they are connected through the same network as well (Table 1).

After analyzing the documents that describe the current operation of IT service for the case study and the problem logs reported by the users, the researcher was able to model the current mode of operation and recommend changes to them based on the ITIL guidelines.

3.1 Service Desk

3.1.1 Current Practice

A support hotline was established that uses a Single Point of Contact (SPOC) for all incidents, as follows:

3.1.2 Problems

- In Fig. 3, the general users follow the reporting path whereas the senior users by-pass it. The incident will not be logged and the communication among the supporting team members may also break down.
- The general users are not well aware of the scope of hotline support. Often, some out-of-scope incidents are not served by the support hotline, and some incidents are

Table 1 Scopes of the IT services provided to the case study

Services	Description
Application support	Provision of application support services on all matters related to the application systems, e.g. answering phone, fax, e-mail, written request and so on.
System maintenance	Bug fixing
	Minor system modification and minor data conversion
	Problem diagnosis
	Documentation update
System monitoring and optimization	Periodic system performance monitoring and tuning on the application system
Production support and ad-hoc processing requests	Liaise with relevant parties to collect and analyze user requirements
	Perform data extraction
	Answer the enquiries on the system data
Environment and operation support	Perform backup and recovery if needed
	Assist system software upgrade and patches
Procurement support	Provide support and advice
	For capacity planning
	On potential technology substitution and cost estimations
	On hardware/software installation and relocation
Planning, drill test and support for disaster recovery and business resumption	Conduct annual disaster recovery and business resumption drill
	Assist in resumption of business and application in case of disaster
Project management and reports	Prepare relevant papers and minutes to management for advice, approval and endorsement
	Prepare periodic progress reports for system performance and achievement
	Coordinate and attend project related meetings
	Prepare agenda, minute and other related document

not requested that need of customers are not delivered to the support team. IT infrastructure and scope of services should enable its users to customize their expectations. Any suggestions for enhancement should be encouraged for improving and maintaining a right balance among people, processes and technology.

3.1.3 Recommendations to Benchmark the Case Study Practice with ITIL

Based on the ITIL framework, recommendations to the existing system following guidelines of ITIL framework have been provided. Guidelines (GL) of ITIL framework for service desk are provided in Fig. 4.

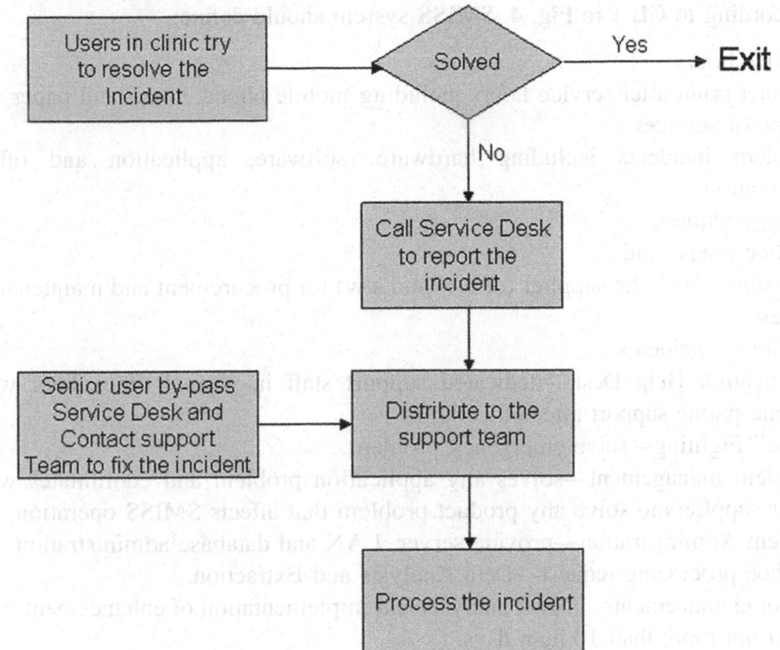

Fig. 3 Current practice of service desk

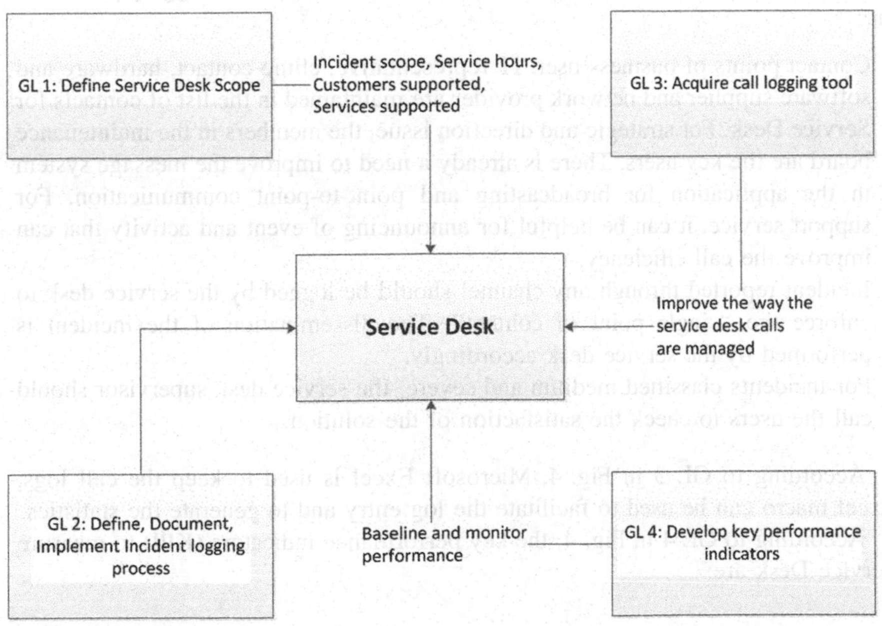

Fig. 4 Service desk guidelines

According to GL 1 in Fig. 4, SMISS system should define:

- Service hours
- Contact point after service hours including mobile phone, email and paper
- Scope of services
- Problem incidents including hardware, software, application and office automation,
- Change request,
- Ad-hoc query, and
- Coordinate with the supplier (of h/w and s/w) for procurement and maintenance issues
- Nature of incidents
- Application Help Desk—dedicated support staff in normal office hour with mobile phone support after office hour.
- "Fire" Fighting—solve emergency problem.
- Problem management—solves any application problem and coordinates with other suppliers to solve any product problem that affects SMISS operation.
- System Administration—provide server, LAN and database administration.
- Ad-hoc processing request—Data Analysis and Extraction.
- Minor enhancement—impact analysis and implementation of enhancement with effort not more than 10 man days.
- Management Activities—proved change, capacity, availability and service continuity management.

According to GL 2 in Fig. 4, the recommendations to the call logging procedure are:

- Contact points of business user, IT representative, clinic contact, hardware and software supplier and network provider are maintained in the list of contacts for Service Desk. For strategic and direction issue, the members in the maintenance board are the key users. There is already a need to improve the message system in the application for broadcasting and point-to-point communication. For support service, it can be helpful for announcing of event and activity that can improve the call efficiency.
- Incident reported through any channel should be logged by the service desk to enforce the "single point of contact". The dissemination of the incident is performed by the service desk accordingly.
- For incidents classified medium and severe, the service desk supervisor should call the users to check the satisfaction of the solution.

According to GL 3 in Fig. 4, Microsoft Excel is used to keep the call logs. Excel macro can be used to facilitate the log entry and to generate the statistics.

According to GL 4 in Fig. 4, the key performance indicators (KPI) to measure Service Desk are:

- Time to log the incident to the incident log database for calls via email, phone, or voice mail,
- Time to acknowledge the user,
- Time to categorize and prioritize the incident,
- Time to start the resolving action,
- Time to complete the action, and
- Percentage of number of satisfaction over the number of medium and high priority incidents.

3.2 Incident Management

3.2.1 Current Practice

Incident handling procedure was established to handle incidents. (Refer to Fig. 5)

3.2.2 Problems

- The logging information was not enough to measure the performance against the server level requirement. There was no escalation procedure defined and support team performance was not measured by any key performance indicator.
- Each service desk staff actually maintained a separate log. The incidents were discussed and prioritized by the Change Advisory Board, but other users were not able to learn the status of the incident being reported.

3.2.3 Recommendations to Benchmark Case Study Practice with ITIL

Based on the ITIL framework, recommendations to the existing system following guidelines of ITIL framework have been provided.

According to GL 1 in Fig. 6, SMISS system should define:

- Maintain centralized database for incident log using the Excel
- Content of the incident log should include
- Unique identity number
- Report date and time
- Log date and time
- Type of call (written, phone, voice message or verbal)
- Nature of incident (enhancement, ad-hoc request, hardware, software, network, application)
- Acknowledgement date and time
- Priority
- Time to return to office in case of non office hour

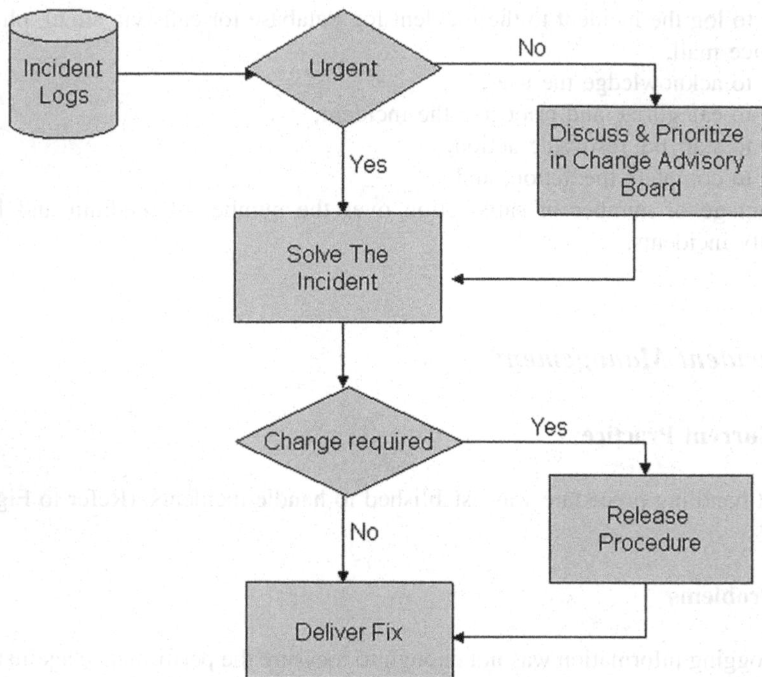

Fig. 5 Current practice of incident management

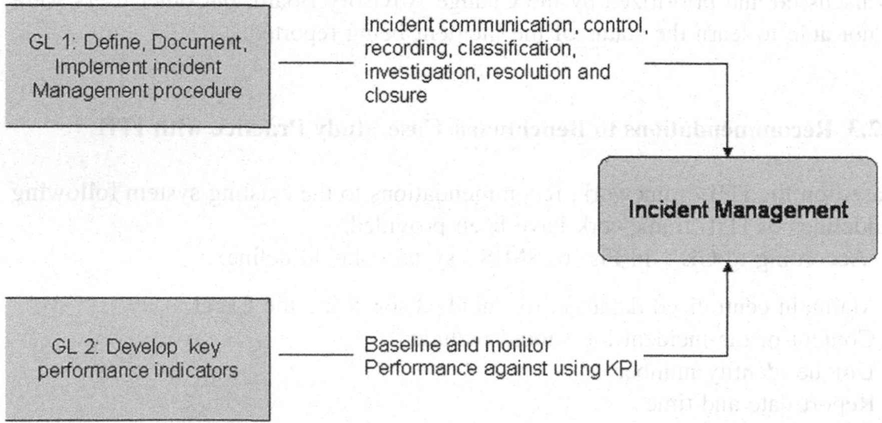

Fig. 6 Incident management guidelines

- Date and time to determine priority of the incident
- Date and time of written reply
- Date and time of analysis result completed
- Date and time of resolution

- Resolution
- Date and time of third party complete the case (resolution is calling the third party)
- Down time and number of workstations affected
- Staff codes perform the receiving, logging, acknowledgement and resolving the incident
- Unique identity number of the configuration item
- Effort estimation
- Effort spent
- Type of the incident
- Application (custom developed programs)
- Hardware
- Software (for example, operating system or system software)
- Network
- Ad-hoc query
- Enhancement
- Query about office automation tool
- Query about application usage
- Other
- Priority of the incident
- Urgent (complete as soon as possible)
- High (complete in 3 days)
- Medium (complete in 2 weeks)
- Low (complete in 2 months)
- Escalation procedure by reporting to support team manager if the incident cannot be solved within the period defined.
- Incidents log is posted to the Intranet site so that users are able to inquire the status of the incident in the log.

According to GL 2 in Fig. 6, the followings are the KPIs for Incident Management:

- number of incidents in open state,
- number of incidents reported within the month,
- number of incidents solved within the month, and
- number of incidents in closing state.

3.3 Change Management

3.3.1 Current Practice

The change management procedure which was established, addresses any change requests required as a result of incident logs, as follows. (Refer to Fig. 7)

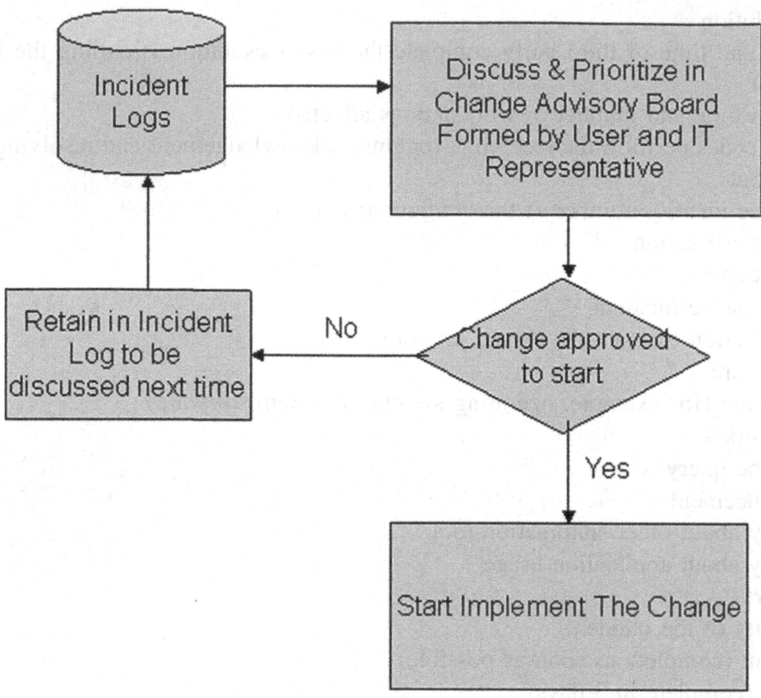

Fig. 7 Current practice of change management

3.3.2 Problems

- The maintenance board doesn't have representative from all functional groups which means, some decisions cannot be made effectively and efficiently.
- The procedure for issuing Request For Change (RFC) is specified; however the duplicated and unpractical requests were not filtered before passing to the Change Advisory Board (CAB). The impact analysis could estimate the effort and the scheduled delay for implementation; however such impact analysis was not conducted.
- There were no key performance indicators for measuring the changes in the system performance.

3.3.3 Recommendations to Benchmark Case Study Practice with ITIL

Based on the ITIL framework, recommendations to the existing system, following the guidelines of ITIL framework have been provided.

According to GL1 in Fig. 8, SMISS system should define scope of CAB:

- CAB should be composed of representative from IT department and staff from each clinic. The agenda and incidents to be discussed will be distributed before

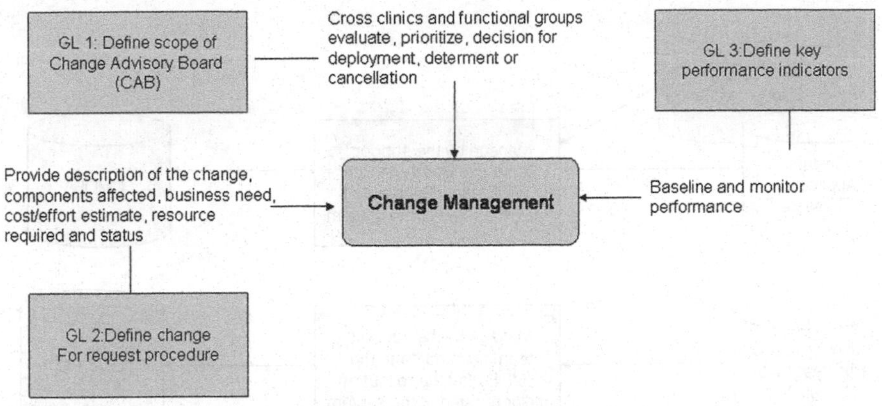

Fig. 8 Change management guidelines

CAB meeting and each functional group should arrange a representative to attend the meeting. The CAB will make decisions for deployment, further analysis, approval or cancellation of changes.

According to GL2 in Fig. 6, the procedure of Request for Change (RFC) should be:

- Incidents that need change should be reviewed before the CAB meeting. Any duplication and unnecessary incidents should be filtered. The status of the filtered incidents will be distributed to the CAB members and the requestor.
- The board should analyze the technical and business impact of the request. The analysis result should be assessed by the CAB.
- If the man-days required exceed the scheduled limit (5 man-days) for the service that will affect the normal support service, then the CAB should determine whether to acquire extra budget for the request or to do it with support team resource but it has lower priority than the service request.
- Change for request should be issued after CAB approves the request. The priority of change request should be high, medium or low. The rollout schedule should also be determined by the CAB. The rollout schedule should be documented and distributed.

According to GL3 in Fig. 6, the followings are the KPIs for Change Management:

- number of failed changes implemented,
- number of emergency changes implemented,
- number of occurrences of the process being circumvented,
- percentages of these numbers, and
- critical level of percentage to be defined and it should be escalated once the level is reached (Fig. 9).

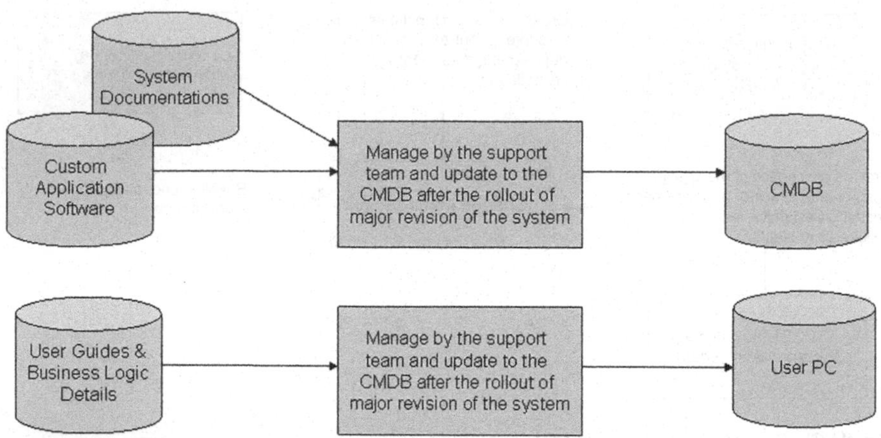

Fig. 9 Current practice of configuration management

3.4 Configuration Management

3.4.1 Current Practice

Problems

- Hardware item was not included in the database since the configuration information of hardware was not available.
- Change Management Database (CMDB) and user prepared materials were not stored centrally and not shared by related parties, and the materials prepared between the parties were not related.
- There were no change records to the configuration items (CIs). The accuracy of the CMDB was not ensured.

3.4.2 Recommendations to Benchmark Case Study Practice with ITIL

Based on the ITIL framework, recommendations to the existing system, following guidelines of ITIL framework have been provided.

According to GL 1 in Fig. 10, SMISS system should prepare configuration management planning:

- Identify the configuration items
  - Hardware—workstation, monitor, printers, external disk, tape, uninterrupted power supply, server, printer, bar code scanner, Chinese input device and rack.

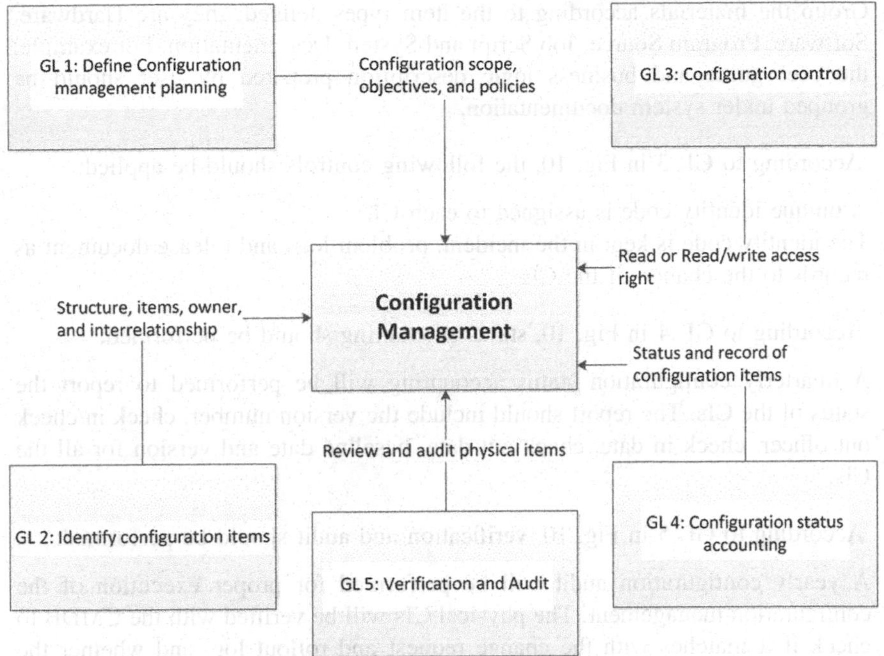

Fig. 10 Configuration management guidelines

- Software—operating system, database server, development tool, version control software. These are the software developed by other vendors for general use.
- Program source—all UNIX and Windows programs custom developed for the customer that the support service is being provided.
- Job script—job scripts to apply change to the production environment.
- System documentation—meeting minute and agenda, incident log, feasibility study, proposal, project plan, analysis & design, system specification, program specification, operation manual, test plan and result, acceptance, approval forms of change implementation, and impact analysis.

- Deployed version control tool to manage the softcopy configuration items. The Software, Program Source, Job Script and System Documentations are stored and protected in a Definitive Software Library (DSL). Standardized configurations of Hardware are stored in the Definitive Hardware Store (DHS).
- CMDB is made ready in the Intranet site to enable user inquiry.

According to GL 2 in Fig. 10, SMISS system should establish the followings configuration structure:

- Identity the owner of the CIs.
- Grant different access right to Read or Read/Write of the CIs.

- Group the materials according to the item types defined; they are Hardware, Software, Program Source, Job Script and System Documentation. For example, the user guide and business logic description prepared by user should be grouped under system documentation.

According to GL 3 in Fig. 10, the following controls should be applied:

- A unique identity code is assigned to each CI.
- The identity code is kept in the incident, problem logs and release document as records to the change of the CIs

According to GL 4 in Fig. 10, status accounting should be performed:

- A quarterly configuration status accounting will be performed to report the status of the CIs. The report should include the version number, check in/check out officer, check in date, check out date, baseline date and version for all the CIs.

According to GL 5 in Fig. 10, verification and audit should be performed:

- A yearly configuration audit will be performed for proper execution of the configuration management. The physical CIs will be verified with the CMDB to check if it matches with the change request and rollout log, and whether the items in the version are all included. Then the configuration activities will be verified to see if all planned activities are conducted accordingly.
- For hardware and software, a yearly audit will be performed to check the labeling and the information regarding the inventory is correct and matches with the information in the CMDB (Fig. 11).

3.5 Release Management

3.5.1 Current Practice

Problems

- There was no release policy; and the support team performed as many changes as possible for each release. Usually the low priority RFCs were left outstanding.
- There was no policy for Hardware and System software upgrades.
- There was no communication to the users about the changes in each release.
- There was no fallback plan in case of unsuccessful releases. There was no plan to merge the emergency fixes and into normal releases.
- Distribution and installation of new releases were error prone since every workstation has to be installed separately.

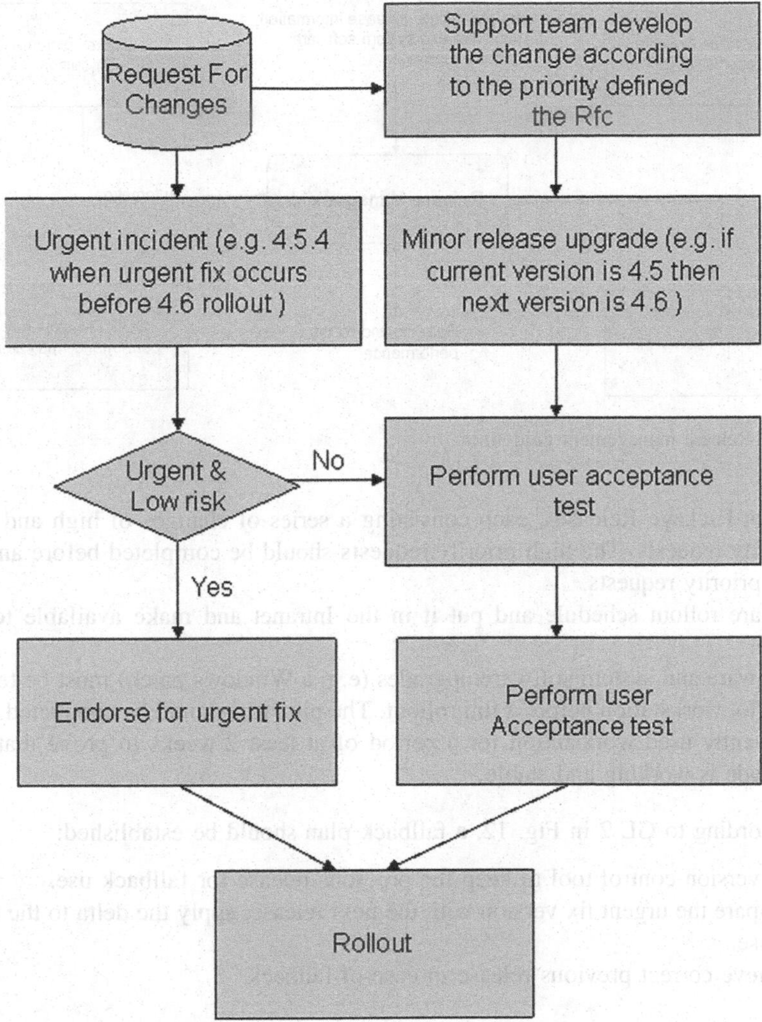

Fig. 11 Current Practice of Release Management

3.5.2 Recommendations to Benchmark the Case Study Practice with ITIL

Based on the ITIL framework, recommendations to the existing system following guidelines of ITIL framework have been provided.

According to GL 1 in Fig. 12, SMISS system should have the following release policy:

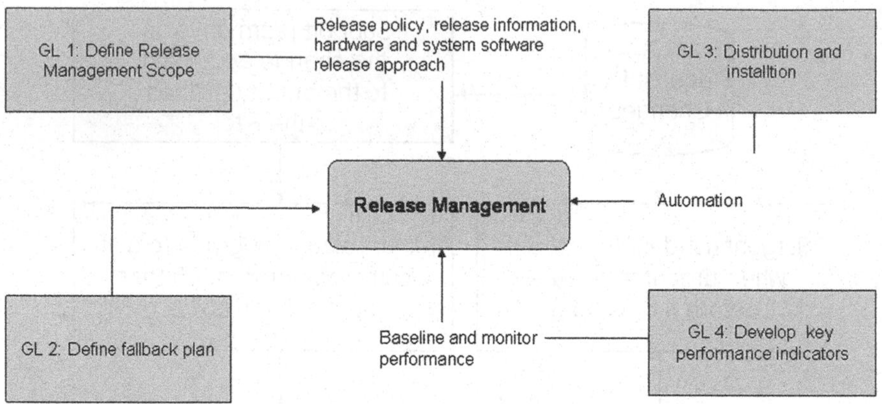

Fig. 12 Release management guidelines

- Adopt Package Releases, each consisting a series of changes of high and low priority requests. The high priority requests should be completed before any of low priority requests.
- Prepare rollout schedule and put it in the Intranet and make available to all users.
- Hardware and system software upgrades (e.g. a Windows patch) must be tested on pilot workstation before a full rollout. The pilot test should be conducted on a frequently used workstation for a period of at least 2 weeks to prove that the upgrade is working and stable.

According to GL 2 in Fig. 12, a fallback plan should be established:

- Use version control tool to keep the previous release for fallback use.
- Compare the urgent fix version with the next release; apply the delta to the next release.
- Retrieve correct previous release in case of fallback.

According to GL 3 in Fig. 12, distribution and installation should be enhanced:

- Inform user of the release and the content in 2 days in advance.
- Automate the distribution and installation by developing an auto-installation module to the system. The module upgrades the workstation module once a new version is found.

According to GL 4 in Fig. 12, the followings are the KPIs for Incident Management:

- number of problem incidents caused, and
- number of occurrences of the process being circumvented.

3.6 Problem Management

3.6.1 Current Practice

The current practice of Service Management Information Support System (SMISS) was mainly reactive, i.e. the support team solved the reported incidents. There was no procedure defined for proactive problem management (PM).

3.6.2 Problems

- The number of incidents was not reduced and the system could not be stabilized.
- Users stopped reporting the repeated incidents and restarted the system to solve the incidents. User satisfaction dropped and blamed the system informally.
- Support team prepared the data extraction manually for each clinic and repeated periodically. User data services effort was not reduced.

3.6.3 Recommendations to Benchmark Case Study Practice with ITIL

Based on the ITIL framework, recommendations to the existing system following guidelines of ITIL framework have been provided.

According to GL 1 in Fig. 13, SMISS system should define a reactive problem management, as follows:

- Conduct monthly review incidents should identify chronic problems by verifying the number of occurrences of the same or similar incidents.
- Build a problem log database using a unique reference number for each problem. This number is updated to the incidents log and incident number should be updated to the problem log for cross reference. It is possible to have multiple incidents pointing to the same problem. The problem log should have the following attributes:
  - Problem reference number
  - Date and time of creation
  - Date and time of solution
  - Created by
  - Solved by
  - Major type (hardware, software or network)
  - Minor type (for example servers, workstation, MS word or router)
  - Supplier (for example, Microsoft or CISCO)
  - Description of problem
  - Incident numbers

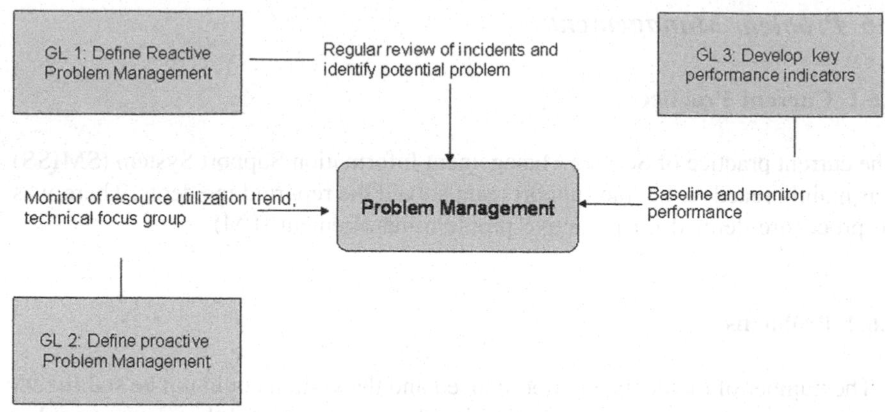

Fig. 13 Problem management guidelines

According to GL 2 in Fig. 13, SMISS system should define a proactive problem management. In general, the problems that are related to the network and CPU, could gradually downgrade the system performance. Users report incidents related to system performance when it becomes unacceptable. The resource utilization trend should be monitored to determine if the system (network or CPU) performance has fallen below an acceptable threshold.

- Build in-house technical focus groups. A focus group for SMISS could monitor:
  - Windows,
  - Unix,
  - Web,
  - Development tool, or a
  - Database.
- A focus group will be able to solve any technical incidents more efficiently in a proactive manner. The focus group should keep the support team informed of any possible problems that could occur in a timely manner.

According to GL 3 in Fig. 13, the KPIs are:

- Number of incidents,
- Average number of incidents related to a problem, and
- Number of problems.

4 ITIL Framework Implementation, Testing and Results

SMISS is a nursing information system developed with the Microsoft Visual Studio development tools. It runs under the Windows operating system; and the workstations are distributed in a local area network. In July 2010, a group of SMISS users and IT representatives were invited to discuss the implementation of ITIL practice as a case study for improving the service. At least one user from each clinic was invited to participate in this discussion (Fig. 14).

Microsoft provides package guidance called Microsoft Operation Framework (MOF) that enables organizations to achieve mission-critical system reliability, availability, supportability, and manageability of IT solutions that are built with Microsoft technologies. To achieve the operations excellence, Microsoft combines the ITIL best practices into MOF, and extends MOF to follow the ITIL code of practice. The MOF provides assessment templates with a set of questions with yes/no answers. Operation guidelines are provided to help users to answer these questions. The questionnaire concerning various performance criteria was prepared using the MOF assessment template. Because the questionnaire has been used before as a successful tool to measure the level of effectiveness of the IT services in an organization according to the MOF guidelines, the answers that were collected from the selected group of users can represent as an important test tool for the system.

All ITIL functions and processes are tested in this chapter except the Financial Management because the simplified data communication and Sharing (SDCS) adopts the financial processes according to the Government practice. At the same period of this research, there was a security audit process conducted by a third party vendor, as well. Most of the users in the test group were also participated in the security audit process to measure the level of effectiveness of Security Management.

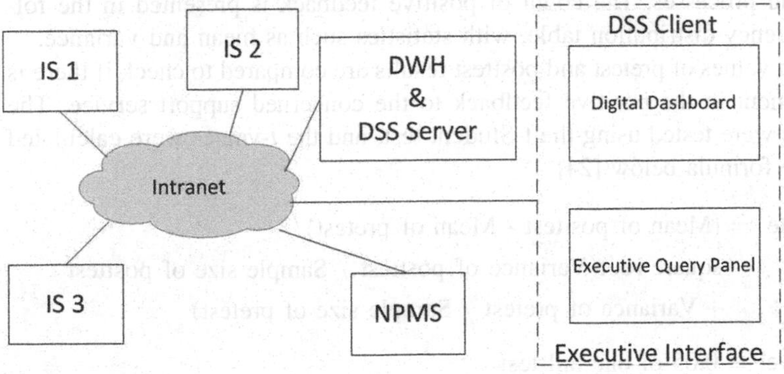

Fig. 14 Decision support system

Table 2 Frequency distribution of service desk result

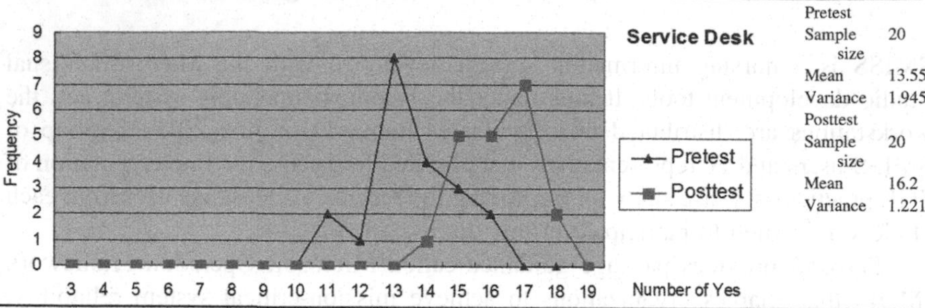

Table 3 Frequency distribution of incident management

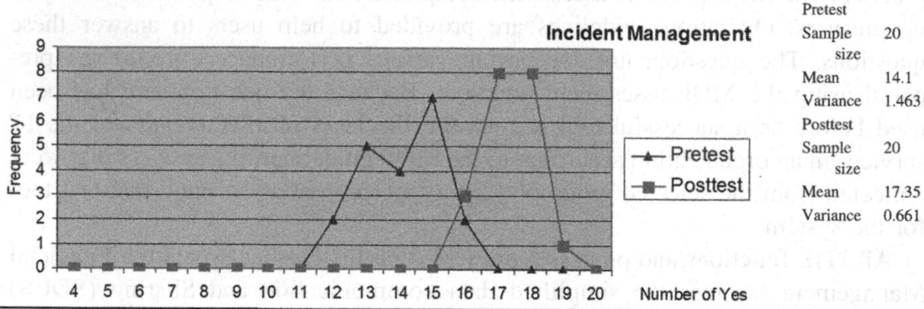

In November 2010, the same group of SMISS users and IT representatives were invited to participate in the research to collect the posttest data. The answers to pretest and posttest data using the same questionnaire are presented in Tables 2 and 3, respectively. The questionnaire was formatted in such a way that the ITIL functions and processes. The count of positive feedback is presented in the following frequency distribution table, with statistics such as mean and variance.

The mean values of pretest and posttest results are compared to check if there is an improvement in the positive feedback to the concerned support service. The mean values were tested using the t-Student test, and the *t*-values were calculated by using the formula below [24]

$$\text{t - value} = (\text{Mean of posttest - Mean of pretest})/ \\ \text{square root}(\text{Variance of posttest} / \text{Sample size of posttest} \\ + \text{Variance of pretest} / \text{Sample size of pretest})$$

Alpha level = 0.05 of one tail test

Degree of freedom = sample size of pretest + sample size of posttest—2 = 38

According t-distribution significance table, the critical value is 1.684 for one tail test.

Null Hypothesis—there is no difference between the pretest and posttest sample means for each of ITIL function and processes.

The table presented the percentage of Yes, before and after implementing ITIL framework which indicates the effect of ITIL and the improvement achieved. From the above table, the null hypothesis of no difference between pretest and posttest mean values for PM, SLM and SM are accepted, which implies that means there is no change in the positive feedback from the test group after the ITIL practices are implemented. The null hypothesis of the other ITIL processes is rejected since the mean values of posttest are larger than the mean values of pretest. It can be concluded that the test group shows more positive feedback after the ITIL practices are implemented (Tables 4–13).

The *t*-value analysis further concludes whether the improvement is significant or not. Then, the processes can be grouped as follows:

Significantly improved—D, IM, CnM, ChM, RM, CAP, AVM and Sec Not improved—PM, SLM and SM

Table 4 frequency distribution of problem management

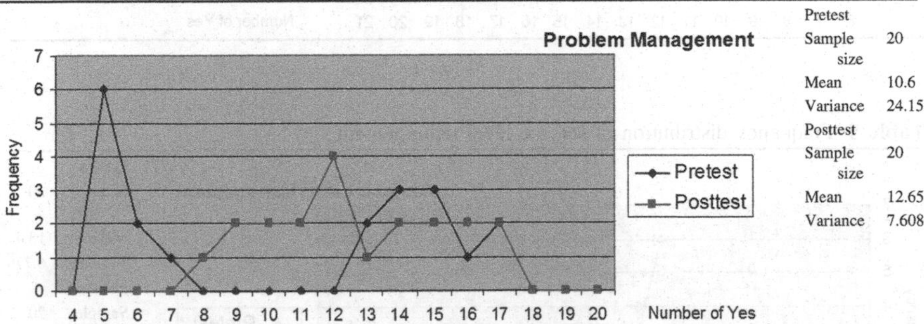

Table 5 Frequency distribution of configuration management

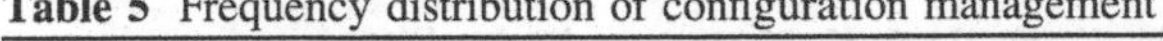

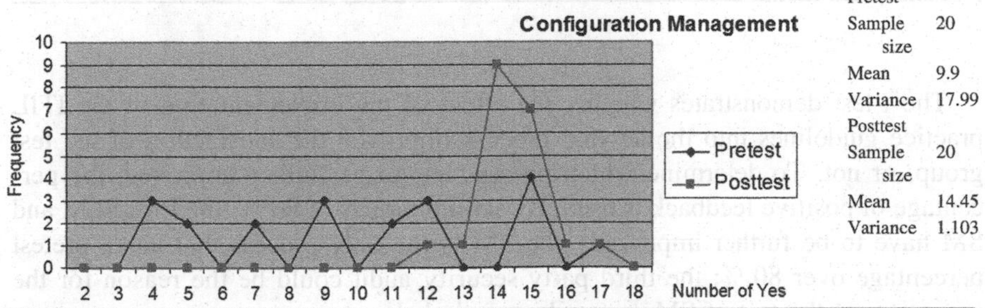

Table 6 Frequency distribution of change management

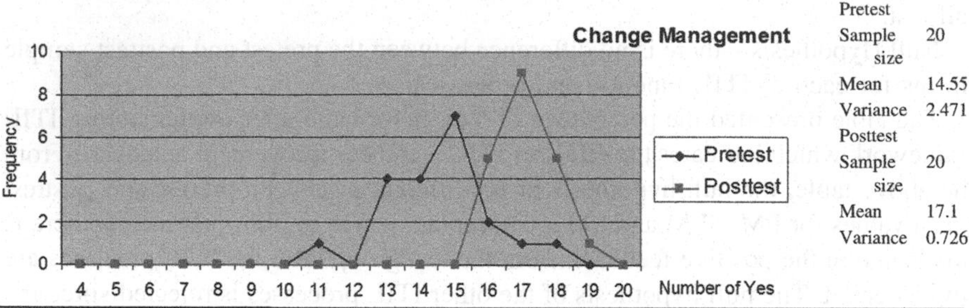

Table 7 Frequency distribution of release management

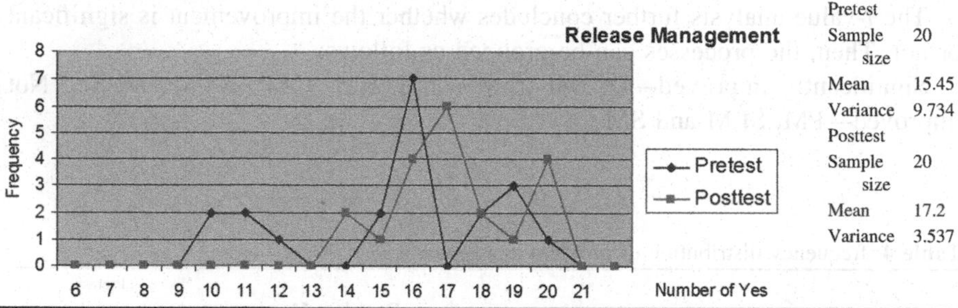

Table 8 Frequency distribution of service level management

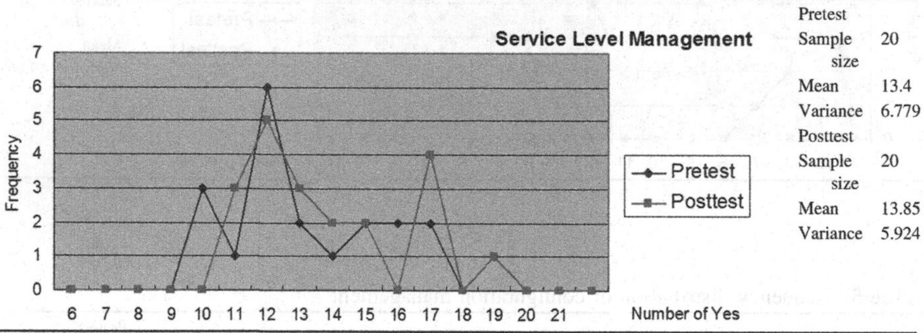

The t-test demonstrates whether the effect of the implementation of the ITIL practice guidelines into the service process improved the satisfaction of the test group, or not. To determine which process has to be further improved, the percentage of positive feedback is used. By setting targets of 80 %, the PM, SLM and SM have to be further improved. The SM is the only process that has a pretest percentage over 80 %; the third party security audit could be the reason for the scenario and the test of SM cannot be concluded.

Table 9 Frequency distribution of capacity management

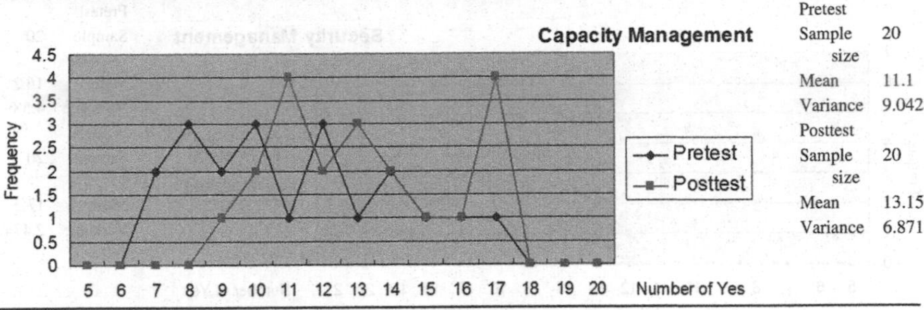

Table 10 frequency distribution of IT service continuity management

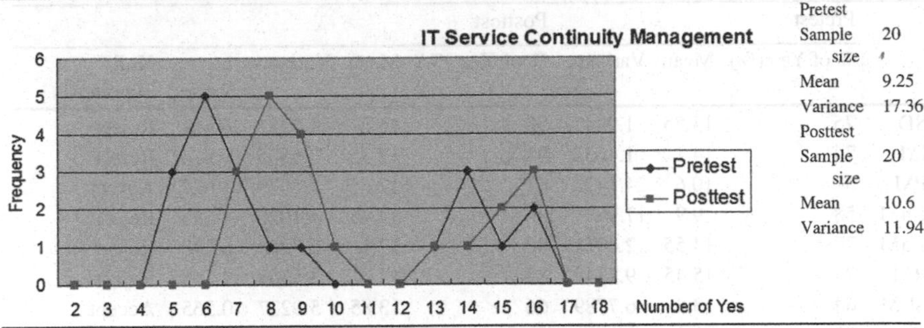

Table 11 Frequency distribution of availability management

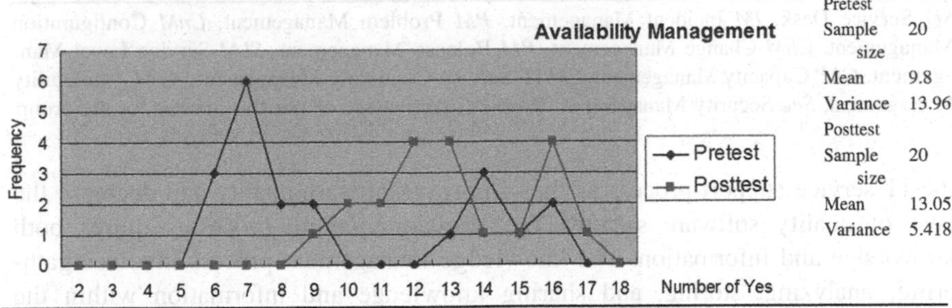

5 DSS Interface to IT Service Support

The ITIL provides a framework for operations and infrastructure while the CMMI (capability maturity model integration) provides a set of improvement goals and a point of reference for appraising current processes. Both CMMI and ITIL improve

Table 12 Frequency distribution of security management

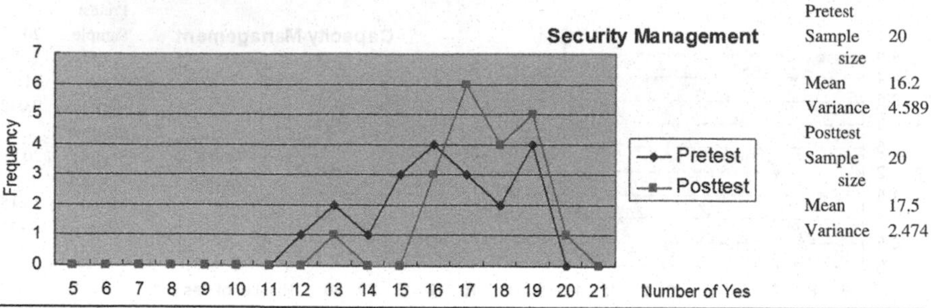

Table 13 Null hypothesis of the services

	Pretest			Posttest				
	% of Yes (%)	Mean	Variance	% of Yes (%)	Mean	Variance	*t*-value	Null hypothesis
SD	75	13.55	1.9447	90	16.2	1.2211	6.661	Reject
IM	74	14.1	1.4632	91	17.35	0.6605	9.974	Reject
PM	56	10.6	24.147	67	12.65	7.6079	1.627	Accept
CnM	58	9.9	17.989	85	14.45	1.1026	4.657	Reject
ChM	77	14.55	2.4711	90	17.1	0.7263	6.378	Reject
RM	74	15.45	9.7342	82	17.2	3.5368	2.148	Reject
SLM	64	13.4	6.7789	66	13.85	5.9237	0.565	Accept
CAP	58	11.1	9.0421	69	13.15	6.8711	2.298	Reject
SM	54	9.25	17.355	62	10.6	11.937	1.116	Accept
AVM	58	9.8	13.958	77	13.05	5.4184	3.302	Reject
Sec	81	16.2	4.5895	88	17.5	2.4737	2.188	Reject

SD Service Desk, *IM* Incident Management, *PM* Problem Management, *CnM* Configuration Management, *ChM* Change Management, *RM* Release Management, *SLM* Service Level Management, *CAP* Capacity Management, *SM* IT Service Continuity Management, *AVM* Availability Management, *Sec* Security Management, *% of Yes* percentage of positive answer for the group

the IT service support process as they improve software quality and decrease the cost of quality software system. The decision making process requires both knowledge and information. The knowledge management process involves gathering, analyzing, storing, and sharing knowledge and information within the organization [28]. Information provides clues to resolve an uncertainty and complexity of an issue, while the knowledge helps in understanding the ambiguity around the issues. A decision support system aids in decision making under the conditions of uncertainty and complexity [29].

The IT support process reengineering is an ongoing process, which requires a continuous monitoring of the KPIs at an operational level and tactical level. The various targets such as green light, yellow light, and red light signals can be established for each KPI. Recent advancements in the telecommunications and computer networking technologies are able to connect any distant and disparate systems together, making it possible to control a remote system from anywhere, based on decisions made in effective management of IT service support process. A decision support system continuously monitors the KPIs and signals appropriate actions that can be performed on any remote system as depicted in Fig. 14.

The software components of Intranet and the information systems (IS) connected to it, are managed by the network performance management system (NPMS). A data warehouse (DWH) system is also connected to the network that extracts, transforms and loads (ETL) all needed data related to the KPIs of IT service support process. The Decision Support Server (DSS) again interfaces with the DWH builds the KPIs of IT service support process and displays them on a digital dashboard of an IT executive responsible for supporting all IT services. Both DWH and DSS server can be housed on the same hardware platform for simplicity or on different systems that are connected together. An IT executive who manages IT service support process runs a DSS client that provides a realtime digital dashboard with all KPIs and alarms suggesting IT service actions. Furthermore, the executive can also perform queries for addional information if needed. The proposed DSS application futher improves the IT service support process and serves as tool for an effective on-going reengineering of IT service support process.

6 Conclusions

Although the results of all KPIs examined in this case study have demonstrated some improvement, it did not fully meet our initial expectations, as some of the processes did not have significant improvement. There are two major possible explanations for this outcome pattern. Firstly, the duration of the test is not long enough for the test group to experience ample improvement. For example, there were no major in SM upgrades performed during the test period. Secondly, some of the intended process reengineering efforts couldn't be fully implemented during the period of this case study, as it required more time, effort, and budget. For example, the Problem Management required a focus group and a pool of technical expertise that was not possible during the pilot project. However, a case study like this one serves as a trigger for major reengineering of business processes. It could motivate the senior management to allocate appropriate budget, and plan a gradual implementation of process reengineering. The ITIL framework consists of a well evaluated, explored and maintained set of guidelines. It certainly serves a tool for exploring process reengineering and improvements while meeting the budget constraints. The case study required a lot of coordination and consensus while identifying process improvements, establishing a process reengineering methodology, and constructing questionnaires for process evaluation.

The experience gained in a case study like this one can alleviate the possibility of expensive mistakes if a major process reengineering is initiated at once.

Actually, the customer company appreciated the efforts in this case study, well received, and motivated for further reengineering of companywide processes. The chapter also proposed a comprehensive DSS client/server system which further improves the IT service support process in reading real time KPIs and IT service actions. Further work to this research can focus on automatic implementation of IT service support actions based on DSS signals.

References

1. Niessink, F., Vliet, H.V.: The Vrije Universiteit IT service capability maturity model. Technical report IR-463, Release L2-1.0. Vrije Universiteit Amsterdam (1999)
2. CCTA (The UK Central Computer and Telecommunications Agency): Information Technology Infrastructure Library, HSMO Books, London, UK (2011)
3. Despont-Gros, C., Mueller, H., Lovis, C.: Evaluating user interactions with clinical information systems: a model based on human-computer interaction models. J. Biomed. Inf. **38**, 244–255, (2005)
4. Benbasat, I., Goldstein, D., Mead, M.: The Case Research Strategy in Studies of Information Systems. MIS Q. **4**, 368–386 (1987)
5. Orlikowski, W., Baroudi, J.: Studying information technology in organizations: research approaches and assumptions. Information Systems Research (1991)
6. Yin, R.K.: Case study research-design and methods. In: Applied social research methods series, vol. 5, 2nd edn. Sage, Newbury Park (1994)
7. Soy, S.: The case study as a research method. Available on the Internet (http://www.gslis.utexas.edu/~ssoy/usesusers/l391d1b.htm). 15 March 2003
8. Ku, S., Suh, Y.-H., Gheorghe, T.: Building an intelligent business process reengineering system: a case-based approach. Intel. Syst. Account. Finance Manag. **5**, 25–39
9. Grigori, D., Casati, F., Castellanos, M., Dayal U., Sayal M., Shan M.-C.: Business process intelligence. Comput. Ind. **53**, 321–343 (2004) http://www.sciencedirect.com, science@direct
10. van der Aalst, W.M.P., Weijters, A.J.M.M.: Process mining: a research agenda. Comput. Ind. **53**, 231–244. (2004) http://www.sciencedirect.com, science@direct
11. Tan Wenan, Shen Weiming, Zhou Bosheng, Li Ling: A business process intelligence system for engineering process performance management. IEEE Trans. Syst. Man Cybern. Part C Appl. Rev. **38**(6), 745–756 (2008)
12. Brenner M., Radistic, I., Schollmeyer, M.: A case-driven methodology for applying the MNM service model. In: Proceedings of the 8th International IFIP/IEEE Network Operations and Management Symposium, 2002
13. van der Aalst, W.M.P.: Business Process Simulation for Operational Decision Support. In: BPM 2007 Workshops, LNCS 4928, pp. 66–77 (2008)
14. Xin, H.: IT service support process meta-modeling based on ITIL. In: International Conference on Data Storage and Data Engineering (DSDE), 2010, pp. 127–131
15. Lahtela, A., Jantti, M., Kaukola, J.: Implementing an ITIL-based IT service management measurement system. In: International Conference on Digital Society (ICDS), 2010, pp. 249–254
16. Jantti, M.: Improving incident management processes in two IT service provider companies. In: 22nd International Workshop on Database and Expert Systems Applications (DEXA), 2011, pp. 26–30
17. Brenner, M.: Classifying ITIL processes; a taxonomy under tool support aspects. In: IEEE/IFIP International Workshop on Business-Driven IT Management, pp. 19–28, 2006

18. Pereira, R., Mira da Silva M.: A maturity model for implementing ITIL V3 in practice. In: 15th IEEE International Enterprise Distributed Object Computing Conference Workshops (EDOCW), 2011, pp. 259–268
19. Lucio-Nieto, T., Colomo-Palacios, R.: ITIL and the creation of a Service Management Office (SMO): a new challenge for IT professionals: an exploratory study of Latin American companies. In: 7th Iberian Conference on Information Systems and Technologies (CISTI), pp. 1–6, (2012)
20. Spremic, M., Zmirak, Z., Kraljevic, K.: IT and Business process performance management: Case study of ITIL implementation in Finance Service Industry. In: 30th International Conference on Information Technology Interfaces (ITI), 2008, pp. 243–250
21. Stake, R.E.: The Art of Case Study Research. Sage, California (1995)
22. Denzin, N.: The research act. Prentice Hall, Englewood Cliffs (1984) (IT Governance Institute, 2000, COBIT 3rd edition network, TM, London)
23. Bell, J.: Doing Your Research Project. Open University Press, Milton Keynes (1992)
24. Zikmund, W.G.: Business Research Methods, 6th edn. The Dryden Press, London (2000) (Forth Worth)
25. Sekaran, V.: Research Methods for Business: A Skill Building Approach. Wiley, New York (1992)
26. Neuman, L.W.: Social research methods: qualitative and quantitative approaches, 3rd edn. Allyn and Bacon, Boston (1997)
27. Sommer, B., Sommer, R.: A Practical Guide to Behavioral Research: Tools and Techniques, 4th edn. Oxford University Press, New York (1997)
28. Phifer, B.: Next-generation process integration : CMMI and ITIL do devops. Cutter IT J. **24**(8), 28–33 (2011)
29. Zack, M.H.: The role of DSS technology in knowledge management. In Proceedings of the IFIP TC8/WG8. 3 International Conference on Decision Support in an Uncertain and Complex World, pp. 861–871 (2004)
30. Caster-Steel, A., Tan, W.: Implementation of IT Infrastructure Library (ITIL) in Australia: Progress and Success factors. In: 2005 IT Governance International Conference, Auckland, New Zealand, 14–15 Nov 2005

Chapter 4
Managing Cloud Services with IT Service Management Practices

Koray Erek, Thorsten Proehl and Ruediger Zarnekow

Abstract Cloud computing represents a fundamentally new trend in the information technology (IT) industry. Especially the practical benefits and the company-specific use cases have become subject of controversial debate. While information systems (IS) play a crucial function in organizations, IT service management (ITSM) is of increasing importance to IT organizations. The need to align IT and business strategy as well as the increase of transparency and quality of IT processes and services are amongst others the main reasons why the IT Infrastructure Library (ITIL)—nowadays the de facto standard of IT service management—was introduced. Against the backdrop of the increasing market and service orientation of IT organizations, the question arises, how cloud computing can affect existing ITSM processes in general, and, which particular ITSM processes are used to ensure an effective management of globally distributed services. This chapter will provide answers to these and related questions.

Keywords IT service management · Cloud computing · Cloud service management portfolio · ITIL

K. Erek (✉) · T. Proehl · R. Zarnekow
Chair of Information and Communication Management,
Institute of Technology and Management, Technische Universitaet Berlin,
Strasse des 17, Juni 135 10623 Berlin, Germany
e-mail: koray.erek@tu-berlin.de

T. Proehl
e-mail: t.proehl@tu-berlin.de

R. Zarnekow
e-mail: ruediger.zarnekow@tu-berlin.de

M. Mora et al. (eds.), *Engineering and Management of IT-based Service Systems*,
Intelligent Systems Reference Library 55, DOI: 10.1007/978-3-642-39928-2_4,

1 Introduction

For small and medium sized enterprises there is a wide range of public cloud services that is available [1]. Contrary to that, large enterprises often prefer closed private cloud solutions, especially to meet compliance policies and to avoid loss of control [2–5]. For business units a cost-effective IT operation and a high degree of flexibility are particularly important [6, 7]. Due to a paradigm shift from a traditional IT department towards a more customer and service centric orientation, the entire management of IT services is becoming increasingly important. For the in-house IT organization and its sequential control within IT service management (ITSM), the IT Infrastructure Library (ITIL) framework has established itself as the de facto standard, consisting of a collection of best practices [8]. This has several reasons but particularly the supply of guidelines for the provisioning of IT services and for processes that should support the business units are the main drivers behind the ITIL adoption [9].

Many corporations can be considered unbiased and generally interested in the subject of cloud computing [3, 10]. The purchase of cloud services can lead to fundamental process changes within ITSM. Continuing this line of thought, it is even possible that the cloud provider could become responsible for all ITSM processes. Depending on the extent of cloud usage on the user side, the process-related activities of the internal IT organization is subject to changes, ranging from small to very large. In this context we attain the following questions, which will be addressed in this chapter:

- How will the significance of ITSM processes in cloud computing change using the example of ITIL on the customer's side?
- What potential for changes do existing ITIL processes have for the cloud computing user?
- What are relevant key performance indicators (KPIs) for managing the cloud services?

For further consideration, it is assumed that both the service provider and the customer have implemented an ITSM based on ITIL (Fig. 1). However, in this chapter the effect on ITIL processes will be considered only in regard to cloud service usage. Market studies show that cloud services are mainly used to complement existing solutions and internal systems; we therefore assume that the purchasing company only partially sources its IT services and resources from the cloud [11].

Decision support systems play a crucial role in ITSM. A corresponding system in ITSM ensures an efficient and resilient flow of service management processes. Event Management might use a neural network to identify, categorize or prioritize events, whereas dashboards are used to visualize Service Level Agreements (SLAs) as well as the current status of the Service Level Targets (SLTs). Rule-based approaches support Capacity and Availability Management functions.

In the next section, the fundamentals of cloud computing and ITSM will be introduced. It is followed by a chapter that combines ITSM and cloud computing,

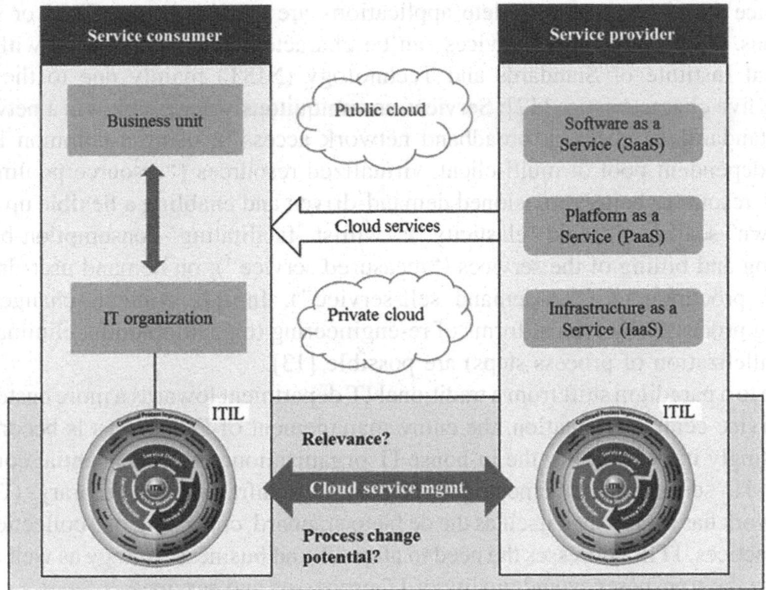

Fig. 1 Usage scenario Cloud Services and ITIL

showing the implications of the cloud computing scenario for an ITSM-oriented IT organization. Furthermore, a detailed reflection of various ITIL processes will be given with the emphasis on changes and shifts (between IT organization and cloud service provider) of ITIL processes. Subsequently, a cloud service management portfolio will be presented and discussed. Thus, the expected contribution of this chapter is a sensitization of the mentioned issues involving cloud computing and ITSM. Moreover, this chapter highlights possible ITSM-related process and organizational changes and will lay open gaps that future research will need to address.

2 Fundamentals

Cloud computing has emerged as one of the most discussed and written about recent IT trend for many organizations. It represents a collection of IT services and resources that can be offered flexible and scalable over the Internet, without requiring a long-term capital lockup and IT know-how. Depending on the depth of vertical integration, the customer can purchase complete software applications as well as the necessary IT infrastructure. Infrastructure as a Service (IaaS) represents the "lowest" layer; the customer receives a flexible IT infrastructure which is scalable both upwards and downwards. On top of that layer, Platform as a Service (PaaS) offers platforms that provide interfaces for cloud infrastructure and tools for the development of cloud applications. The top layer is represented by Software as a Service (SaaS), where complete applications are provided, e.g. CRM or ERP solutions. Cloud computing services can be characterized in accordance with the National Institute of Standards and Technology (NIST) mainly due to the following five characteristics [12]: Services are ubiquitously accessible via a network by a standardized access ("broadband network access"); using a common location-independent pool of multi-client, virtualized resources ("resource pooling"); with IT resources being provisioned demand-driven and enabling a flexible up and/or down scaling ("rapid elasticity"); whilst facilitating consumption-based metering and billing of the services ("measured service"); on demand user-driven service procurement ("on-demand self-service"). In this context, changes in existing processes by typical forms of re-engineering (e.g. automation, elimination or parallelization of process steps) are possible [13].

Due to a paradigm shift from a traditional IT department towards a more customer and service centric orientation, the entire management of IT services is becoming increasingly important. For the in-house IT organization and its sequential control within IT service management (ITSM), the IT Infrastructure Library (ITIL) framework has established itself as the de facto standard, consisting of a collection of best practices. ITIL addresses the need to align IT and business strategy as well as to increase the transparency and quality of IT processes and services.

ITIL V3 was published in 2007, and in 2011, an update was introduced. The ITIL 2011 Edition contains no change in the basic concept of the ITIL Service Lifecycle; however, there are many improvements in terms of general consistence and clarity [14, 15]. The Service Lifecycle consists of five phases, whereas each phase contains several processes. The Service Strategy phase deals with the development of a strategy to serve customer needs. Designing IT services is goal of subsequent phase called Service Design. The corresponding processes address the design of new and change of existing services. Next, ITIL Service Transition builds and deploys these services, while Service Operation ensures that the services are delivered effectively and efficiently. The Incident and Problem Management processes are located in Service Operation phase. Finally, the processes of Continual Service Improvement conduct improvement initiatives like customer survey or improvement plans. This chapter takes ITIL 2011 Edition as a basis and makes use of the corresponding processes.

Besides ITIL, there are some other ITSM frameworks like MOF [16] and HP IT Service Management Reference Model [17]. This chapter considers only ITIL because of its wide distribution.

3 IT Service Management for Cloud Services

In the following paragraphs relevance and change potential for the five phases of the ITIL service lifecycle will be discussed against the background of cloud computing. For this purpose, a closer look will be taken at the ITIL processes, which are undergoing significant changes and/or have a high significance for cloud service management.

The **Service Portfolio Management** (2) is responsible for managing the service portfolio and ensures an adequate composition of IT services. Within the scope of cloud portfolios there are well-defined modules (services), so that a flexible orchestration is possible [18]. Here, a large change potential is contingent upon the so-called self-services that are being made available to the service consumers automatically and independent from the location. The service portfolio is also influenced by the requirements of the business units in which cloud services had been suggested. Consequently, a fast update of the portfolio, which meets the requirements of the business unit, is possible. A holistic monitoring of the service usage is difficult for cloud services because the possibilities for the IT organization to influence said services are limited. In particular the resulting consequences for the migration and the shutdown of services present companies with new challenges ("Retired Services"). Questions regarding data storage, archiving and deletion (e.g. the IaaS service Dropbox) have to be addressed (Table 1).

The **Financial Management for IT Services** (3) is accountable for the financial planning, analysis, reporting and billing of IT services [18, 19]. Cloud services get deployed precisely under the premise to yield cost savings [4]. A transparent breakdown of expenses is provided by the method of consumption-based billing in the cloud, so that business divisions only have to pay for the actual usage of the resources [20]. The internal billing of self-booked cloud services (e.g. the SaaS service Sales Cloud by Sales Force) requires that the accounting information between provider and customer is exchanged electronically (Electronic Data Interchange for Administration, Commerce and Transport—EDIFACT). Thus far-reaching process changes have to take place to enable flexible and short-termed billing of cloud computing services. For the model of the community cloud (sharing of resources by the members), it is necessary to define billing mechanisms to ensure a consolidated cost allocation for the communities' virtual pool of resources (Table 2).

The **Business Relationship Management** (5) maintains customer relationships, conducts surveys regarding customer satisfaction and is generally responsible for the business unit's demands. Because of minor restrictions that apply when using

Table 1 Service portfolio management—KPI and tool for managing cloud services

KPI	Planned new services
	Percentage of new services which are developed or introduced following strategic reviews
	Unplanned new services
	Percentage of new services which are developed or introduced without being triggered by strategic reviews
Tool	Axios assyst
Tool type	IT service management tool
Interface	Web service

Table 2 Financial management for IT services—KPI and tool for managing cloud services

KPI	Adherence to approved budget
	Percentage of IT expenses exceeding the approved budget
	Total actual versus budgeted costs
	Total actual costs as a percentage of budgeted costs. Calculated for an entire service portfolio
Tool	Axios assyst
Tool type	IT service management tool
Interface	Web service/EDIFACT

cloud services (e.g. firewall settings), it is possible to access external data and applications outside of the closed system. That raises the question to what extent the IT organization supports the cloud services or, according to internal corporate compliance requirements, rigorously rejects those. Additionally, it is increasingly simpler for business units to order IT services directly, regarding cloud computing. This leads to the development of many small island solutions in the organization, for both applications and devices. However, this cannot be supported by the IT organization or be adapted to the necessary business requirements. This phenomenon (also known as "Shadow IT") is a challenge for many IT organizations because it complicates the alignment with the requirements of the business units [21]. Transparent communication between the IT organization and the business unit is possible through clearly defined service packages in the cloud, benefiting from a high customer orientation and integration (Table 3).

The **Service Catalogue Management** (7) designs, develops and maintains the service catalog of the IT organization. The importance of the service catalog increases through the use of cloud services, as it is the central planning tool and can counteract "Shadow IT" [18]. Ideally, the service level agreements (SLAs) of the cloud services are transferred automatically (import interface/standardized data exchange), resulting in an average process change potential. Especially the automated information gathering leads to challenges for the service consumers as cloud services are standardized and can only be accessed through existing interfaces according to the cloud level. In cases in which information can be read automatically the information content is limited to basic, mostly technical data (latency, state of the virtual machine, the number of users, etc.) [22]. A validation

Table 3 Business relationship management—KPI and tool for managing cloud services

KPI	Customer complaints
	Number of received customer complaints
	Customer satisfaction per service
	Average measured customer satisfaction for each service (including standard deviation), determined by means of customer satisfaction surveys
Tool	LimeSurvey
Tool type	Survey tools/web-based survey
Interface	API

Table 4 Service Catalogue Management—KPI and tool for managing cloud services

KPI	Used services
	Number of services recorded and managed within the service catalogue as a percentage of those being delivered and transitioned in the live environment
	Service description mismatch
	Number of variances detected between the information contained within the service catalogue and the deliver services
Tool	OTRS—service catalogues management component
Tool type	IT service management tool
Interface	Import interface/standardized data exchange

of SLAs or locating where data is processed is not possible. In addition to that, monitoring of cloud services in distributed and global usage scenarios is much more complex (Table 4).

In the **Service Level Management (8)** explicit and formally defined terms are negotiated with the business units. With sourcing of cloud services, all parameters of the service design, such as availability and capacity, have to be included in the contracts and monitored in the process. KPIs for applications should be considered in contracts (end-to-end service levels), since they have a greater significance [19, 23]. For (public) cloud services, there are usually only standardized SLAs that come into use [5], which are passed to the business unit through the Service Level Management. To ensure adequate quality of service, it is important to monitor the SLAs and to determine management responsibilities for cloud services, ensuring that the expertise and the authority to implement service changes are given. In cloud computing, it becomes increasingly important to control and monitor multiple providers and their services. Due to the variety of cloud providers, a SLA dashboard seems plausible [18] respectively the adaptation of the processes in terms of multi-vendor strategies (Table 5).

Capacity Management (9) ensures that the capacities of the IT services and infrastructure are sufficient to provide the goals (performance and capacity)

Table 5 Service level management—KPI and tool for managing cloud services

KPI	Services covered by underpinning contracts
	Number of services where SLAs are backed up by corresponding underpinning contracts
	Underpinning contract breaches
	Number of underpinning contract breaches
Tool	Nagios
	Icinga
	OpenNMS
	Shinken
Tool type	Computer system, network, and infrastructure monitoring with focus on availability as well as capacity aspects
Interface	Different protocols such as SNMP, WMI, HP-iLO, SSH, HTTP, SMTP

Table 6 Capacity Management—KPI and tool for managing cloud services

KPI	Incidents due to capacity shortages
	Number of incidents occurring because of insufficient service, component or cloud capacity
	Exactness of capacity forecast
	Deviation of the predicted capacity development from actual course
Tool	Nagios
	Icinga
	OpenNMS
	Shinken
	VMware vSphere
	Microsoft Operations Manager
	Amazon CloudWatch
Tool type	Computer system, network, and infrastructure monitoring with focus on capacity aspects, panning components of virtualization platforms, and cloud provider tools
Interface	Protocols such as SNMP, WMI, HP-iLO, SSH, HTTP and web service

economically. In the cloud scenario, the process is limited to determining (support for the identification of appropriate cloud providers) and monitoring capacities (for a feedback to the Financial Management). For cloud services, an inaccurate determination of demand can be corrected with the flexible increase of capacities (elasticity). Ideally, the cloud services scale automatically, resulting in a high degree of process change potential because activities from the traditional Capacity Management process are no longer required. The importance of the process will remain mostly the same, since a monitoring of the capacities will still be necessary. For example, using Amazon Web Services (AWS), missing or redundant virtual machines can be booked or switched off within a few minutes. Monitoring, however, remains the responsibility of the service consumer who, in this context, often can receive support tools from the provider (e.g. Amazon CloudWatch) (Table 6).

The **IT Service Continuity Management (11)** manages risks and implements mechanisms for ensuring continuity, to the extent that the minimal requirements agreed in SLAs can be achieved in case of extraordinary events. In the context of cloud services, it is necessary to fixate these contractual requirements and to draw up a contingency plan with the involvement of alternative providers. In this case, restrictions on the interoperability of cloud providers (potential lock-in effect) are a problem which should not be underestimated [1]. Because of more complex structures (lack of interoperability or distributed resource pools) the development of contingency plans is becoming an increasingly significant challenge to face. Through the loss of control by the service consumer when using cloud services, emergency management is only possible to a limited extent. In case of storage services in the cloud, a redundant usage respectively recovery by the customer is possible. A cloud ERP system (e.g. the SaaS solution BusinessByDesign by SAP), however leaves the customer with no scope of action when it fails, resulting in loss or non-availability of data (Table 7).

Table 7 IT service continuity management—KPI and tool for managing cloud services

KPI	Business processes with continuity agreements
	Percentage of business processes which are covered by explicit service continuity targets
	Gaps in disaster preparation
	Number of identified gaps in the preparation for disaster events (major threats without any defined counter measures)
Tool	Realtech the guard!
Tool type	Monitoring (capacity, availability, performance) and planning
Interface	Protocols such as SNMP, WMI, HP-iLO, SSH, HTTP and web service

In the cloud computing scenario the **Information Security Management (12)** plays a key role, because the service consumer needs to be protected at all times in terms of confidentiality, integrity and availability [19, 23]. The process plays an important role in the selection of potential cloud providers by passing the safety requirements on to the Supplier Management. The record of safety-related incidents at the cloud provider should—if possible—be handled at this part of the process chain. This particular process has a primarily planning and monitoring character in the context of cloud services. New challenges lie in the monitoring of the external provider and in a complex security concept that considers a location and device independent use. For cloud services, as compared to internal operation or traditional IT outsourcing, only some of the controlling aspects (data encryption, secure transfer protocols, rights management, etc.) are available to the IT organization (Table 8).

The **Supplier Management (13)** is responsible for contracts with suppliers and associated issues. Especially for cloud services, "clean" arrangements between the two parties are necessary so that the increased demand for compliance (services, processes and systems) is addressed adequately [19, 23]. In this scenario the number of traditional negotiations decreases because the contracts are signed electronically (process change). Nevertheless a careful examination of the contracts is of utmost importance to ensure that the cloud provider actually fulfills the inter alia legal (e.g. server location abroad) and organizational requirements. A negotiation of contracts respectively post-negotiation in the traditional sense will most likely only be possible with private cloud providers rather than public

Table 8 Information security management—KPI and tool for managing cloud services

KPI	Security-related service downtimes
	Number of security incidents causing cloud service interruption or reduced availability
	Security tests
	Number of security tests and trainings carried out
Tool	Verinice
	SecuMax
Tool type	Information security management system (ISMS)
Interface	Web service

Table 9 Supplier management—KPI and tool for managing cloud services

KPI	Agreed underpinning contracts
	Percentage of contracts underpinned by Underpinned Contracts
	Identified contract breaches
	Number of contractual obligations which were not fulfilled by cloud providers (identified during contract reviews)
Tool	ServiceNow IT service management suite
Tool type	IT service management tool
Interface	Automatic information exchange with the cloud provider via an interface

cloud providers. This process is a key element because, based on the cloud computing scenario, more and more external cloud services are purchased from different providers (multi-vendor strategies) [19] (Table 9).

The **Event Management** (**21**) monitors the Configuration Items (CIs) and IT services. In addition to that, a filtering of occurring events takes place. While using cloud services the monitoring effort of CIs decreases whereas the monitoring of cloud services in turn increases (proactive monitoring). The network monitoring software Nagios provides capabilities for the monitoring of cloud services such as Amazon Web Services EC2 and S3. Nagios can in turn be, for example connected via an interface to OTRS, an ITSM ticketing system (service consumer). Such measures can only be relayed through a request to the provider, as the IT organization itself does not have access to the IT resources. Especially SaaS solutions are more difficult to monitor due to the high technical levels of abstraction compared to IaaS and PaaS services. Ideally, maintenance (with limit to regard the accessibility), on the part of the cloud provider gets passed on automatically to the service consumer with the result that the event management system does not unnecessarily raise incidents (Table 10).

In **Incident Management** (**22**), all incidents are managed. This process has the primary goal of restoring IT services to the business units as quickly as possible. In the cloud scenario the process is of medium importance because cloud service relevant incidents (cloud incidents) are forwarded to the cloud provider. The transfer

Table 10 Event management—KPI and tool for managing cloud services

KPI	Events
	Percentage of events that occurs in the context of the cloud services
	Self-resolved events
	Percentage of events that occurs in the context of the cloud services and resolves itself
Tool	Nagios
	Icinga
	OpenNMS
	Shinken
Tool type	Monitoring with rule-based or neural network components for event filtering and pattern recognition
Interface	Protocols such as SNMP, WMI, HP-iLO, SSH, HTTP and web service

Table 11 Incident management—KPI and tool for managing cloud services

KPI	Cloud incidents
	Number of incidents that are related to the cloud services
	Cloud incident resolution time
	Average resolution time for cloud incidents
Tool	OTRS—Open-source ticket request system
	RT—Request tracker
Tool type	IT service management tool—help desk software
Interface	Information exchange with the cloud provider via web service

of the cloud incidents may happen through a branch operation in the incident management process. For compliance reasons, the IT organization of the service consumer has to be integrated in this process [19]. In addition to that, self-service portals (OTRS Help Desk) gain in importance (e.g. for creating tickets or restoring passwords) where these portals should ideally have a direct link to the information systems of the cloud provider. The Service Desk has to provide information about the used cloud services when this information is not available as self-service (Table 11).

As part of the **Access Management (24)** users (e.g. from a division) can get authorized to use particular IT services. Additionally, the Access Management executes the requirements received by the Information Security Management. The process has a continuing importance as single sign-on mechanisms between the existing IT services and the various cloud services are difficult or impossible to implement. Moreover, a superior authority that monitors and implements the role management in the cloud needs to stay in place. A process change potential yields different role concepts in a multi-cloud provider environment (Table 12).

The Continual Service Improvement (CSI) is using quality management methods to continually improve the efficiency and effectiveness of services and processes. As part of the **Service Review (26)** recommendations for process optimizations can be made with the goal to increase the quality of the service and to make it more economical. In the **Process Evaluation (27)**, **Definition of CSI Initiatives (28)** and **Monitoring of CSI Initiatives (29)**, initiatives to improve processes and services are defined and monitored. Cloud services must also meet

Table 12 Access management—KPI and tool for managing cloud services

KPI	Granted individual rights
	Number of accounts with individual rights. These accounts didn't fit the standard rights model
	Manual rights assignment
	Number of accounts where a manual rights assignment was necessary and the automatic rights assignment process didn't operate
Tool	Verinice
	LDAP
Tool type	Information security management system (ISMS) or directory service
Interface	Web service

Table 13 Continual service improvement—KPI and tool for managing cloud services

Service review	
KPI	Service reviews
	Number of formal service reviews carried out during the reporting period
	Identified weaknesses
	Number of weaknesses which were identified during service review, to be addressed by improvement initiatives
Process evaluation	
KPI	*Process benchmarkings, maturity assessments, and audits*
	Number of formal process benchmarkings, maturity assessments, and audits carried out during the reporting period
	Process evaluations
	Number of formal service evaluations carried out
Definition and monitoring of CSI initiatives	
KPI	*CSI Initiatives*
	Number of CSI initiatives, resulting from identified weaknesses during service and process evaluation
	Completed CSI Initiatives
	Number of CSI initiatives which were completed during the reporting period
Tool	Nagios, Icinga, OpenNMS or Shinken
	LimeSurvey
	Pentaho or Jaspersoft
	Alfresco
Tool type	Monitoring, survey tools, business intelligence and document management
Interface	Web service

the requirements of continuous improvement (CSI requirements), which should be fixated by contract with the cloud provider [19]. Such a contractual arrangement with private cloud providers seems very possible, contrary to public cloud providers which usually do not given any individual configuration options [5]. In particular, said requirements should contain the interval in which the measurement is executed, analyzed and improved. In this case, once can safely consider the IT organization as being transparent and it passes the requirements of the business unit on to the cloud provider [19]. The CSI processes will retain the same relevance as before the introduction of cloud services because even with increasing purchases of cloud services the ITSM processes must be continuously improved (Table 13).

4 ITSM Portfolio for Cloud Services

The increasing demand of cloud services in IT organizations can lead to a significant change in the relevance of ITIL processes for service consumers. At the same time, a major potential for process change is created through the introduction

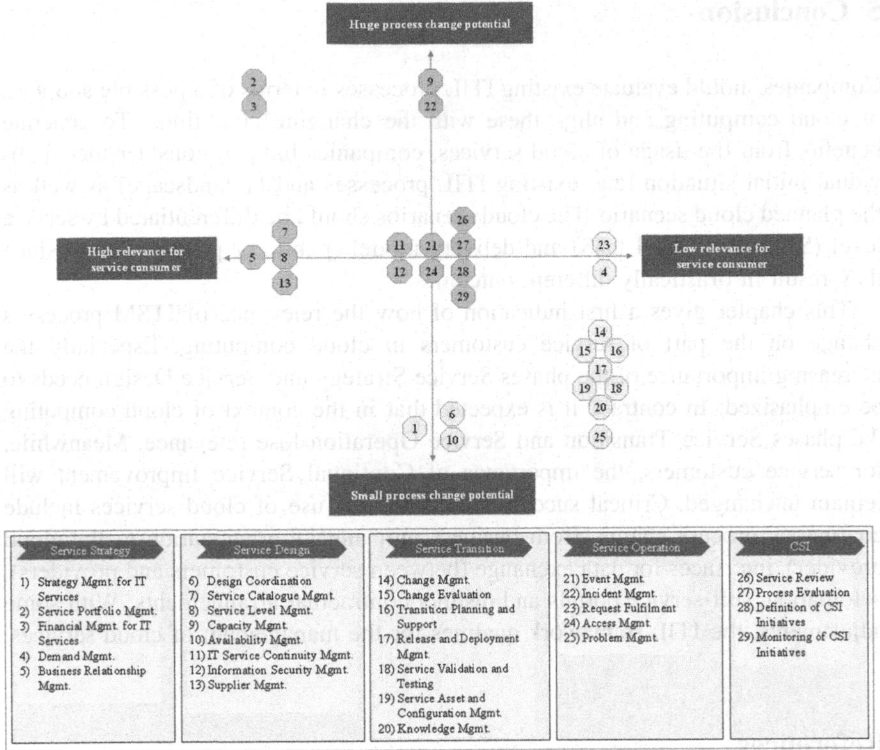

Fig. 2 Cloud service management portfolio illustration

of cloud services. There are considerable differences between the traditional services and cloud services that need to be taken into account. This can be measured by including the five NIST characteristics, which notably influence the process change potential of the ITSM processes.

Companies that use cloud services have to expect process changes in the context of IT service management, especially with regard to the relevance and process operation (Fig. 2). It is foreseeable that the phases Service Design and Service Strategy will gain in importance (dark gray colored processes). Due to high levels of standardization and automation of cloud services the operation (Service Operation) and transfer (Service Transition) of IT services will simplify and could thus become less important. For the service consumer of cloud computing it will be increasingly important to plan the deployment of services and strategically align those with the corporate objectives (see [24]). Particularly the Service Strategy phase has to utilize the individually different potentials of cloud computing and derive the correct conclusions from the challenges. As an overarching process for permanent improvement, the Continual Service Improvement (CSI) is (in the context of cloud computing) crucial for the service recipient. However, due to the outsourcing of tasks and resources, the range of influence in CSI declines.

5 Conclusion

Companies should evaluate existing ITIL processes in terms of a possible adoption of cloud computing and align these with the changing conditions. To generate benefits from the usage of cloud services, companies have to consider their individual initial situation (e.g. existing ITIL processes and IT landscape) as well as the planned cloud scenario. The cloud scenarios should be differentiated by service level (SaaS, PaaS, and IaaS) and delivery model (public or private cloud), since they result in drastically different outcomes.

This chapter gives a first indication of how the relevance of ITSM processes change on the part of service customers in cloud computing. Especially the increasing importance of the phases Service Strategy and Service Design needs to be emphasized. In contrast, it is expected that in the context of cloud computing the phases Service Transition and Service Operation lose relevance. Meanwhile, for service customers, the importance of Continual Service Improvement will remain unchanged. Critical success factors for the use of cloud services include continuous process chains (from business unit and IT organization to the cloud provider), interfaces for data exchange (between service customers and providers), automation, self-service portals and distinct contractual arrangements. With some adjustments, the ITIL framework qualifies for the management of cloud services.

References

1. Repschläger, J., Zarnekow, R., Wind, S., Turowski, K.: Cloud requirement framework. requirements and evalutation criteria to adopt cloud solutions. In: European Conference on Information Systems (ECIS) (2012)
2. Geczy, P., Izumi, N., Hasida, K.: Cloudsourcing: Managing cloud adoption. Global J. Bus. Res. **6**(2), 57–70 (2012)
3. KPMG: Cloud monitor 2012. Eine Studie von KPMG durchgeführt von PAC (2012)
4. Marston, S., Li, Z., Bandyopadhyay, S., Zhang, J., Ghalsasi, A.: Cloud computing—The business perspective. Decis. Support Syst. **51**(1), 176–189 (2011)
5. BITKOM: Cloud Computing—Was Entscheider wissen müssen. Ein ganzheitlicher Blick über die Technik hinaus. Positionierung, Vertragsrecht, Datenschutz, Informationssicherheit, Compliance. http://www.bitkom.org/files/documents/BITKOM_Leitfaden_Cloud_Computing-Was_Entscheider_wissen_muessen.pdf (2010)
6. Armbrust, M., Fox, A., Griffith, R., Joseph, A.D., Katz, R.H., Konwinski, A., Lee, G., Patterson, D.A., Rabkin, A., Stoica, I., Zaharia, M.: Above the clouds: A Berkeley View of Cloud Computing. http://www.eecs.berkeley.edu/Pubs/TechRpts/2009/EECS-2009-28.html (2009)
7. Rawal, A.: Adoption of cloud computing in India. J. Technol. Manag. Grow. Econ. **2**(2), 65–78 (2011)
8. Bause, M.: ITIL und Cloud Computing: Welchen Mehrwert bietet ITIL mit seinem Service Lifecycle-Ansatz für Cloud Computing. Whitepaper (2009)
9. APMG: ITIL Glossar und Abkürzung, http://www.itil-officialsite.com/International-Activities/ITILGlossaries_2.aspx (2011)
10. Repschläger, J., Zarnekow, R.: Cloud Computing in der IKT-Branche (2011)

11. Repschläger, J., Zarnekow, R.: Erfolgskritische Faktoren und Kundensegmente im Cloud Computing. Empirische Studie bei kleinen und mittelgroßen Unternehmen in der Informations- und Kommunikationsbranche (2012)
12. Mell, P., Grance, T.: The NIST Definition of Cloud Computing. Recommendations of the National Institute. Special Publication 800-145, Gaithersburg (2011)
13. Davenport, T.H.: Process Innovation - Reengineering Work through Information Technology. Harvard Business School Press, Boston (1993)
14. Best Management Practice: ITIL Lifecycle Suite 2011. One single source for: Service Strategy, Service Design, Service Transition, Service Operation, and Continual Service Improvement. TSO, London (2011)
15. BMP: ITIL Update FAQs—Summer 2011. http://www.best-management-practice.com/gempdf/ITIL_UPdate_FAQs_Summer_2011_June11.pdf (2011)
16. Microsoft Corporation: Microsoft Operations Framework 4.0 core documentation. http://technet.microsoft.com/library/cc506049.aspx (2008)
17. HP: The HP IT Service Management (ITSM) Reference Model. ftp://hp.com/pub/services/itsm/info/itsm_rmwp.pdf (2003)
18. Cannon, D., Wheeldon, D., Lacy, S., Hanna, A.: ITIL Service Strategy. TSO, London (2011)
19. itSMF: Positionspapier cloud computing und IT service management (2010)
20. Fry, M.: 5 questions about ITSM and cloud computing. Whitepaper (2010)
21. Györy, A., Cleven, A., Uebernickel, F., Brenner, W.: Exploring the shadows: IT governance approaches to userdriven innovation. In: European Conference on Information Systems (ECIS) (2012)
22. Labes, S., Stanik, A., Repschläger, J., Kao, O., Zarnekow, R.: Standardization approaches within cloud computing: evaluation of infrastructure as a service architecture. In: Federated Conference on Computer Science and Information Systems (FedCSIS) (2012)
23. Morin, J.-H., Aubert, J., Gateau, B.: Towards cloud computing SLA risk management: issues and challenges. In: Hawaii International Conference on System Sciences (HICSS) (2012)
24. Repschläger, J., Hahn, C., Zarnekow, R.: Handlungsfelder im Cloud Computing: Relevanz und Reifegrade des Cloud Computings in typischen Prozessphasen. Forschungsumfrage. IT Operations Day (2012)

Chapter 5
Towards Semantic Support for Business Process Integration

Roberto Pérez López de Castro, Inty Sáez Mosquera and Jorge Marx Gómez

Abstract In this chapter we attempt to cover the major contributions in the semantic approach to business process integration. We ought to present our perspectives and vision about how to carry out semantic support for business process integration and a set of recommendations based on the major contributions of frameworks, methodologies, procedures and tools, considering open issues as well as challenges in this field; providing some insight into the semantic shifting of Business Process Management (BPM), the current work and future research trends on the topic. There is also presented a proposal for a framework design that combines the semantic supported modeling with the orchestration of integration processes in the Supply Chain Management (SCM) context.

Keywords BPM · Semantic models · ITSM · SCM

1 Introduction

With intensified globalization, the effective management of an organization's business processes became ever more important. Many factors such as the rise in frequency of ordered goods, the need for faster information transfer, quick decision making, the need to adapt to change in demand, more international

R. P. L. de Castro (✉) · I. S. Mosquera
Central University "Marta Abreu" of Las Villas, Las Villas, Cuba
e-mail: robertop@uclv.edu.cu

I. S. Mosquera
e-mail: intysaez@uclv.edu.cu

J. M. Gómez
"Carl von Ossietzky" University of Oldenburg, Oldenburg, Germany
e-mail: jorge.marx.gomez@uni-oldenburg.de

M. Mora et al. (eds.), *Engineering and Management of IT-based Service Systems*,
Intelligent Systems Reference Library 55, DOI: 10.1007/978-3-642-39928-2_5,

competitors and demands for shorter cycle times are challenging the profitability and survival of big and small companies [21].

In a bid to deal with these challenges information technology was given the task to manage business processes. Over recent years, chief information executives have realized the importance of information technology in 'making the difference' by changing business processes. Hence, building process capabilities is of major interest to organizational decision making now and in the future [19]. The motivation to manage its processes comes from the organizations efforts to continuously align their actual business processes, as executed by the diversity of systems in use, with the should-be processes as derived from business needs.

Business Process Management (BPM) is defined by van der Aalst (2003) as "Supporting business processes using methods, techniques and software to design, enact, control and analyze operational processes involving humans, organizations, applications, documents and other sources of information" [22]. Hepp (2005) argues that BPM is "the approach to manage the execution of IT-supported business operations from a business expert's process view rather than from a technical perspective" [9]. Software tools supporting the management of such operational processes became known as Business Process Management Systems (BPMS) [10].

In the second definition Hepp points out the main problem in initial stages of BPM, the divorce between the Business Perspective and the IT Perspective; creating a need for a unified view on business processes in a machine-readable form that allows querying their process spaces by logical expressions, with the lack of such a machine-readable representation of their process space as a whole on a semantic level been a major obstacle towards mechanization of BPM [9]. As it turns out is quite difficult for business analysts to use the existing tools due to high complexity, and equally difficult to IT specialists to understand the business needs and what a process represents.

Recently, Business Process Modeling Notation (BPMN) [4] created by the BPMI group had emerged as a standardized notation for process modeling, joining many other notations e.g. UML ADs, IDEF, ebXML and EPCs. In order to improve the flexibility of business processes, the time between process modeling and the transformation into executable code needs to be reduced. Therefore, Business Process Modeling Languages (BPMLs) must provide execution languages and in turn, offer more explicit notation elements on all perspectives. But a huge number of different elements will be confusing for process description purposes, thus a basic set of elements, like the BPMN offers is also needed.

BPMN aims at bridging the gap between business process design and process implementation. It was to allow for the automatic translation from the graphical process diagram into the Business Process Execution Language (BPEL) representation that may be then executed using a Web services technology. Although the goal of automatic translation is very appealing, the intention failed in practice for a number of reasons. One of them is that BPMN is a graph-oriented language and its mapping to the block-structured BPEL representation is challenging. In addition, BPMN allows designing ill-formed processes that cannot be translated directly into a set of the BPEL executable instructions. Nowadays, the BPMN modeling is supported by many tools. Some of them allow also for the translation from BPMN diagrams to BPEL, but this functionality is neither fully automated nor supported with semantics [16].

One of the ontologies developed within the SUPER project (sBPMN) aims to overcome these drawbacks with the use of semantics [1]. With the creation of semantic descriptions, this ontology can be tested against competency questions, e.g. what are the elements of a given process, what are the sequence flow connection rules, what is the execution order of activities within the process, which objects can be a source of compensation association, how a certain type of activity can be triggered, among others, in order to prove the domain coverage as well as reasoning capabilities.

There are already several attempts to provide semantic support in the different phases of process management cycle; here we discuss some relevant approaches.

In Ref. [23] are discussed potential benefits from adding semantics to Business Process Management from a methodological point of view, with a focus on the Modeling and Configuration phases. Semantic enrichment of business processes can be achieved by representing a business process or part of it, with an ontology-oriented formalism (semantic lifting) and mapping it to reference ontology. De Nicola (2008) shows how it is possible to generate an OWL representation of a BPMN diagram [6]. Also in Ref. [3] they show how an overlap in the conceptual model of OWL-S and BPEL4WS can be used to overcome the lack of semantics in BPEL4WS by mapping BPEL4WS to OWL-S.

The development of middleware for integrating and automating enterprise business processes, advanced transaction models for business processes, and architectures and implementation techniques for business process management systems; has been receiving over the past years a lot of contributions. The growth in e-commerce and the blurring of enterprise boundaries (e.g. Supply Chains), has raised a huge interest in business process coordination, especially for inter-organizational processes, and standards are being introduced by various consortia such as RosettaNet and ebXML. In this work we provide a perspective on technologies for intra and inter enterprise business process coordination.

Developments in frameworks, methodologies, procedures and tools for semantic business process integration has seen a lot of industry specific and technology constrained solutions, however there are several projects in the work which will be relevant to address in the following sections. But despite the fact process modeling languages are developing into allowance of process composition and execution through web services, they are not able until now to define the integration mechanisms for heterogeneous schemas in the data exchange.

This chapter aims at providing some insight into the semantic shifting of BPM, the issues and challenges in regard to business process integration support, the current work and future research trends on the topic; there is also presented a proposal for a framework design that combines the semantic supported modeling with the orchestration of integration processes in the Supply Chain Management (SCM) context.

2 Background

Before exploring the semantic support to business process integration, it is always good to begin with an overview of BPM basics and standards.

To better understand the definitions, concepts and features of BPM, we should start from the BPM Life Cycle. According to [22] the BPM Life Cycle consists of:

1. Process Design—In this stage, business processes are electronically modeled into BPM systems (BPMS). Graphical Standards are dominant in this stage.
2. System Configuration—This stage configures the BPMS and the underlying system infrastructure. This stage is hard to standardize due to the differing IT architectures of different enterprises.
3. Process Enactment—Electronically modeled business processes are deployed in BPMS engines. Execution Standards dominate this stage.
4. Diagnosis—Given appropriate analysis and monitoring tools, the BPM analyst can identify and improve on bottlenecks and potential fraudulent loopholes in the business processes. The tools to do this are represented by Diagnosis Standards (Fig. 1).

BPM standards and specifications are based on established BPM theory and are eventually adopted into software and systems. BPM Standards and BPM Systems are also described sometimes as "BPM-Enabling Technologies".

Ko (2009) categorizes BPM standards by their features and their BPM life cycle perspective. This categorization allows filtering out Web service standards and B2B standards from BPM Standards, and also allows further classifying BPM Standards into Graphical, Execution, Interchange and Diagnosis standards:

1. Graphical Standards: this allows users to express business processes and their possible flows and transitions in a diagrammatic way.
2. Execution Standards: to computerize the deployment and automation of business processes.
3. Interchange Standards: to facilitate portability of data, e.g. the portability of business process designs in different graphical standards across BPMS; different execution standards across disparate BPMS, and the context-less translation of graphical standards to execution standards and vice versa.

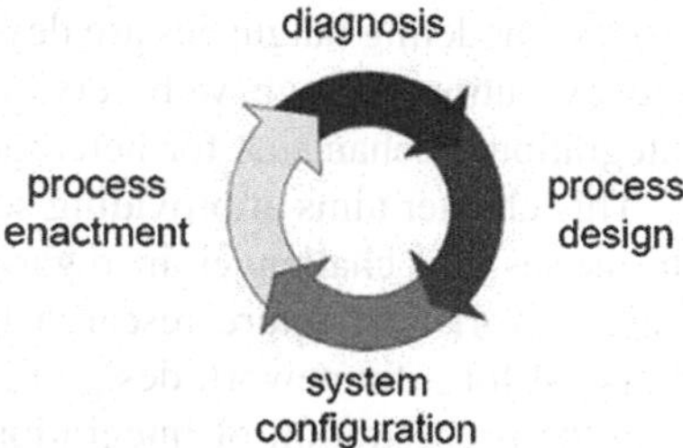

Fig. 1 BPM life cycle [22]

4. Diagnosis Standards: provide administrative and monitoring (such as runtime and post-modeling) capabilities. These standards can identify bottlenecks, audit and query real-time the business processes in a company [11].

The heterogeneity of business process languages is a notorious problem for BPM [15]. Actually the amount of standards available is somehow confusing and problematic sometimes, especially when it comes to the goal of integration. Let's take a look at some of the most prominent ones according to Ko's classification.

2.1 Graphical Standards

Graphical Standards are currently the highest level of expression of business processes (i.e. most natural to human beings) while the lowest level (i.e. the most technical) are the Execution Standards. Even though the Interchange Standards aim to bridge the Graphical Standard to the Execution Standard or vice versa, the translation can sometimes be imperfect, as both Standards are conceptually different [18].

Graphical notations like UML AD and BPMN are easy for non-technical business users to understand and use. Compared to the text based execution-level standards like BPEL, graphical standards visually reveal patterns, loopholes and bottlenecks of a business process. However, the finite set of process diagram elements may restrict design freedom somewhat. The UML AD and the BPMN share many graphical symbols (e.g. rounded rectangles for activities, diamonds for decisions, etc.). These similarities are understandable because both UML AD and BPMN are designed to represent procedural business processes. The differences between BPMN and UML can be understood by considering the intended users of both notations. While BPMN was targeted at business analysts, UML (its Activity Diagram) was primarily targeted for software development. Although the UML 2.0 development upgraded the Activity Diagram to accommodate business analysts, it is still technically oriented [12, 24].

Although graphical standards provide a high-level representation of business processes, its focus is on flow control. Graphical standards are weak on the formulation, evaluation and measurement of the fulfillment of goals. Because of the absence of semantic and computational formalisms in graphical notations, their models were not able to fully translate into executable code. There was always some loss of data or semantics of the control flow.

2.2 Execution Standards

Execution standards enable business process designs to be deployed in BPMS and their instances executed by the BPMS engine. There are currently 2 prominent execution standards: BPML and BPEL [2, 20]. In compare, BPEL is more widely adopted in several prominent software suites (e.g. IBM Websphere, BEA AquaLogic BPM Suite, and SAP Netweaver, among others).

Execution standards are ideal for automation by computers. Syntax-based and block-oriented execution standards facilitate business process automation in IT systems. Currently, many execution standards (e.g. BPEL, BPML, etc.) are based on the well-structured XML. XML is easily customizable and scalable and yet has rich block-oriented features like nesting, structures, and good parsing capability, meaning older versions of business process models can be easily modified without the need for a drastic overhaul. They capture many hidden semantics that graphical standards cannot. With a clear set of syntax, nested features and a formalized method of expressing business processes, execution standards can encapsulate logical details succinctly.

However they are limited by this low level. Like assembly languages in computing, the sequence, activities and the linkages in and between business processes are not visually obvious in execution standards. They require some technical knowledge and web-services know-how. Unlike the flowchart-like graphical standards, the execution standards require some technical knowledge; in recent years, primarily web services. This is a veritable obstacle for process-owners and business analysts who are conversant in business process design but not the technical implementation.

Prior to the industry's recent consolidation towards BPEL, there were several proposals to streamline business process execution. Ko (2009) classifies them as follows:

- Extensions of programming languages: After the ascendancy of workflow management in the late nineties, many programming languages began to cater to the design of web services. A prominent example is the jBPM (Java Business Process Management), which is supported on many Java systems like the community developed JBOSS. Consequently, the problems and limitations of programming languages were directly inherited by business process designs. For example, in order to link with other programming language oriented business process extensions, one will need to build interfaces Web Service based proposals (e.g. APIs), which are often not dynamic, scalable and manageable.
- The other category of execution standards are those based on web services. The web services paradigm focuses on service oriented concepts, meaning that programs or business processes can be created by mixing and matching modular programs each serving a particular function. Service oriented programs are reusable and are very adaptive to changes in requirements and computing environments [11].

Because of the growing dominance of the Service Oriented Architecture (SOA), the web-service based proposals are currently the most influential. Execution standards based on adaptive paradigms are better received due to the user's need to survive in an increasingly globalized business climate.

2.3 Interchange Standards

As mentioned earlier, interchange standards are needed to translate graphical standards to execution standards and exchange business process models between different BPMS's [15]. There are currently two prominent interchange standards: Business Process Definition Metamodel (BPDM) by OMG and XML Process Definition Language (XPDL) by the WfMC.

BPDM is criticized as a complex and user-unfriendly standard. BPDM is relatively immature with no software tool using it, while XPDL is well-accepted and stable, having had a 10-year history. In general interchange standards offer a "globally accepted" file format to save process definitions since business process models in different BPMS are perfectly compatible. But due to fundamental differences in graph-oriented graphical standards and block-oriented execution standards, the quality of transformation of the interchange standards is limited by different syntax and structures. For instance, a cyclical and temporal implication in a graphical standard cannot be easily transformed into an execution standard. The translation of recursive capabilities from an execution standard to a graphical standard is an even more challenging task. Currently in the industry, translation from graphical to execution is easier than that from execution to graphical standards. This applies to XPDL and even BPDM. This limitation raises doubts as to whether the "bridge between the business analyst and the IT specialist" is near in sight. The need to exchange business process models amongst systems is not an everyday activity and XPDL formats sufficient for most purposes. Also, many workflow management systems (and BPMS) already have the capability to translate, in a constricted way, their graphical designs into XML execution codes.

2.4 Diagnosis Standards

The core difference between workflow management and contemporary BPM lies in the diagnosis portion of the BPM life cycle [22]. Diagnosis standards monitor and optimize business processes running in and across companies. Audit trails, real-time business process information, trend analysis and bottleneck identification are just some of the important diagnostic tools. As relevant standards in this area we could mention Business Process Runtime Interface (BPRI) and Business Process Query Language (BPQL). However as this standards do not play a major role in this paper we will limit them to just a mention here.

2.5 Remarks and Discussion

Graphical standards are easily interpreted by business analysts but lack computational formalisms. Execution Standards enable business process automation (i.e. the Process Enactment stage of the BPM Life Cycle). However, they are rather limited in expressing loops and cycles commonly found in real-life business processes. BPEL is beginning to embody many of the capabilities of BPML.

Interchange Standards currently translate Graphical to Execution Standards so business process models in different BPMS are interchangeable. However, the fundamental differences between graphical and execution standards severely limit any translation. Prominent Interchange Standards are XPDL and BPDM. Currently, XPDL is more widely used as BPDM is still a fledgling standard with not many software adaptations. In summary, Graphical and Execution Standards lack computational formalisms whereas Interchange Standards should be scalable and flexible to accommodate new standards and versions.

The industry is currently consolidating towards BPMN as the graphical standard, XPDL as an interchange standard, and BPEL as an execution standard. These 3 standards address the process design and process enactment stages of the BPM Life Cycle.

3 Semantic Support to Business Process Integration

3.1 Business Process Management: Semantic Shifting

Researchers from the Semantic Web community have also identified that the modeling of higher level semantics of business processes are currently limited. By hybridizing Semantic Web Services and Business Process Modeling, [9] highlighted the main semantic limitation of Execution Standards as the lack of machine accessible semantics, and argued that the modeling constructs of Semantic Web Services frameworks are a natural solution to this.

BPM editor tools support modelers in building correct diagrams only from the syntactic point of view. Enriching them with ontologies can bring many advantages as the possibility of applying advanced reasoning techniques, aimed at the identification of contradictions and mistakes in the model specification, and the possibility of organizing BP models repositories, with advanced search and retrieval facilities. Furthermore, semantic technologies can substantially support Business/IT alignment.

With regard to the IT community, the application of ontologies and semantics-based solutions is already a promising reality, since several initiatives are gaining consensus. They are mainly focused on web service discovery and composition (i.e., WSMO, SAWSDL, and OWL-S).

There are several developed and ongoing researches [3, 6, 23] and projects (SUPER, FUSION, e3value) around the topic with the intention to include semantics in the process. There are approaches applying ontologies to describe enterprise models and business processes in general, to show the potential benefits of the application of ontologies for companies; following with the automation of the transition from business process models to monitored execution, and the analysis of what went wrong using the business vocabulary and competency questions that could be delivered by ontologies in the latest steps, with attempts to automate processes using SOA and semantic web services.

The SUPER methodology owns a proper business process "life cycle" that is enriched with the semantic connotation of the overall SUPER framework. Figure 2 shows their definition of the life cycle as a Semantic Business Process (SBP) Life Cycle.

Fantini (2007) describes the Semantic Business Process Life Cycle, organized in four main phases, a top and a bottom layer:

- The SBP Modeling phase, which allows business analysts to obtain process model with ontological annotations, enabling effective semantic search for process model and fragments reuse;

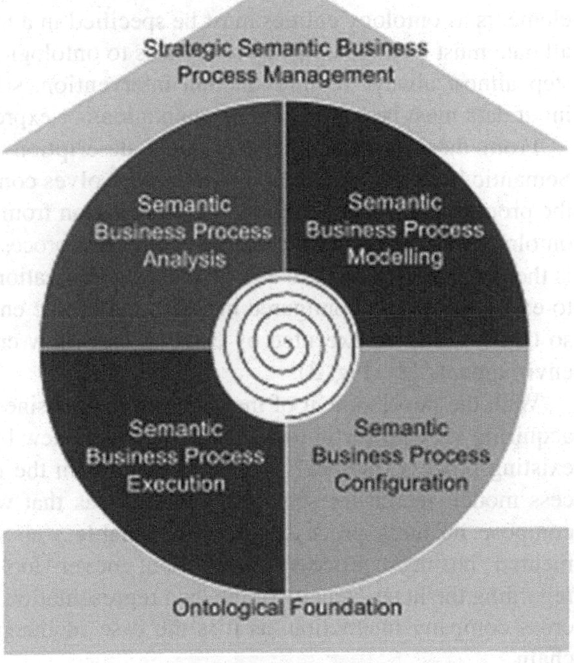

Fig. 2 Semantic business process life cycle [7]

- The SBP Configuration phase, with an efficient semi-automated mapping from business analyst comprehensible to IT-oriented executable specifications (to bridge the Business-IT gap), with the possibility of discovering and composing complex Semantic Web service to match previously defined task goals;
- The SBP Execution phase, enabling the run-time discovery and composition of SWS and thus the highly automated execution of business processes;
- The SBP Analysis phase, to monitor business processes and KPIs and analyzing already executed processes for continuous process improvement;
- A top layer, the Strategic SBP Management, linked to the enterprise strategy and thus the choice of objectives, processes and indicators to be applied in the four SUPER phases;
- A bottom layer, concerning with the ontological foundation of the overall SUPER framework which enables the SUPER components to work thanks to the various ontologies and ontology layers used and developed in the project [7].

Following these guidelines business analyst can use well-known flowchart-like graphics to model a new business process on his computer. Primary input to an SBPM environment can be process models coming from modeling tools, standards, reference process libraries given by ERP vendors, etc.; these process specifications come in two major types of formalisms: standardized process notations, e.g. BPEL or EPC, or proprietary representations. This terminology can be matched to concepts and relations from ontologies; this means the process elements as ontology entities must be specified in a machine-readable manner and all data must be augmented by references to ontologies used in the framework, this step almost always requires human intervention, since the formal semantics of input data must be augmented by annotations or expressed using richer constructs.

From these outputs, the executable description of the process is deployed. Semantic business process configuration involves composition (implementation of the process using web services), then translation from a business process modeling ontology to a semantically enhanced business process execution language, which is then further serialized to an executable specification. Finally, should be possible to export processes configured inside the semantic environment to BPEL or EPCs so that it could be executed by existing workflow engines or EPC based tooling environments [8] (Fig. 3).

With the development of these approaches business managers and analysts are acquiring very powerful tools, they can model new business processes, search for existing process fragments, automatically fill in the missing elements in the process model, search for semantic web services that will deliver the functionality, compose business processes out of available web services and execute implemented business process models; but never looking at the inconsistencies regarding the heterogeneity of the data representation schemas, specially related to cross company interaction, as it is the case of integration processes in a supply chain.

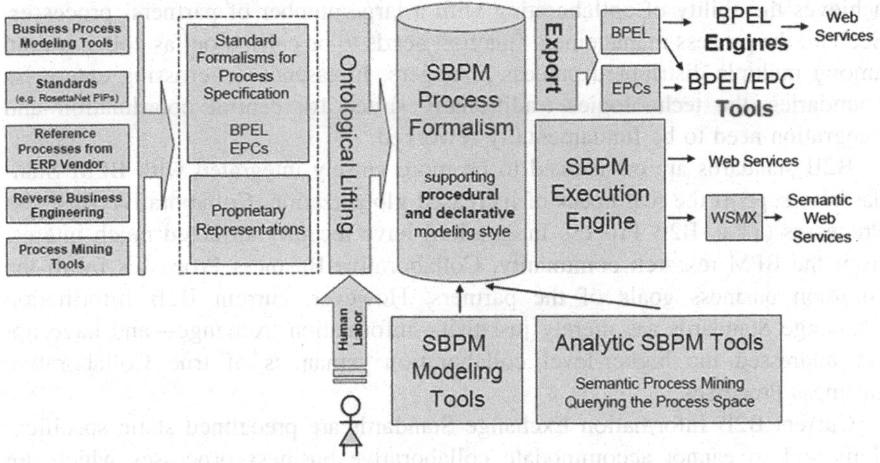

Fig. 3 SBPM environment [8]

3.2 Semantic Support to Business Process Integration, Issues and Challenges

While the management of internal business processes via BPM is crucial, a serious BPM practitioner and researcher should never overlook how organizations collaborate with each other. One must remember companies exist mainly to turn in a profit, via the fundamental activities of buying and selling of products or services, Processes exist within and across companies to support this high-level goal of making profits. Hence, an efficient methodology to support collaborative business processes in B2B collaborations is crucial.

Different forms of middleware have been introduced to enable integration and automation of business processes, both within and across organizations. Message brokers, transactional queue managers, and publish/subscribe mechanisms provide means for automating processes by allowing the component applications to post events and to react to events posted by other component applications. The logic for how to react to events and chain the steps of the process typically resides in the applications themselves. This provides autonomy for the applications and flexibility in that the logic for reacting to events can be changed as business needs or policies change [5].

The potential business value of streamlining inter-enterprise business processes has fueled a renewed interest in process management technologies. However, the conventional intra-enterprise process management architecture faces a number of challenges. First of all, there must be a mechanism to allow participating enterprises to reach agreement on the business process description and the data to be exchanged during process execution. A library of common, standard processes must exist so that by binding to a common, standard process, an enterprise achieves the ability of collaborating with a large number of partners' processes. Second, the process management function needs to be carried out as collaboration among multiple distributed process managers. In essence, in crossing enterprise boundaries, the technologies traditionally suited for central coordination and integration need to be fundamentally reworked.

B2B standards are envisioned to be more closely integrated with BPM Standards to support the real needs of increased globalization. Collaborative Business Processes (a.k.a. B2B Process Integration) have recently attracted much interest from the BPM research community. Collaborative Business Processes fulfill the common business goals of the partners. However, current B2B Information Exchange Standards are merely just that—information exchange—and have not yet addressed the higher-level collaboration semantics of true Collaborative Business Processes.

Current B2B Information Exchange Standards are predefined static specifications and so cannot accommodate collaborative business processes which are dynamically formed. In reality, the identification of suppliers or service providers is still mostly executed manually, even if B2B information exchange standards are available in the market. Furthermore, the establishment of automated standardized B2B information exchange requires a costly setup prior to the actual usage. This means that before fully reaping the benefits and efficiency of the B2B information exchange standards, companies have to evaluate their existing legacy systems, IT infrastructure and licensing schemes prior to utilizing these standards. Currently, B2B Information Exchange standards do not embody contextual information. If some context can be embedded in web-services supporting the underlying methods, a first move to context-aware business process systems will have been made.

3.3 Framework, Methodologies, Procedures and Tools

Developments in frameworks, methodologies, procedures and tools for semantic business process integration has seen a lot of industry specific and technology constrained solutions, however there are several projects in the work which will be relevant to address. In Ref. [14] we can find a comparison of several workflow annotation and composition approaches, including projects SUPER and FUSION.

Preist (2005) uses semantic Web Services technologies for discovery and integration in a logistic Supply Chain [17] and Kotinurmi (2008) presented innovative projects using an ontologically enhanced RosettaNet [13]; but they focus on service discovery and solving the heterogeneity issues in message interchanges, not in process integration.

The major objective of these projects is to "raise Business Process Management (BPM) to the business level, where it belongs, from the IT level where it mostly resides now" [8]. These projects attempt to address the semantic gaps of current web service-based business processes.

But despite the fact process modeling languages are developing into allowance of process composition and execution through web services, they are not able until now to define the integration mechanisms for heterogeneous schemas in the data exchange. For the sake of addressing this problem, we'll present and discuss our proposal for the design of a framework to semantically support the modeling and orchestration of logistic integrated processes, with focus on Supply Chain Management.

The proposed framework combines semantic Business Process Modeling, with support from the SCOR (Supply-Chain Operations Reference-model) ontology; this will allow modeling the logistic integration processes in the Supply Chain based on a recognized standard. The main goal is to define the mechanisms to orchestrate the integration processes in the Supply Chain and provide the business analysts in charge of modeling the new processes common knowledge base to assist the proceeding (Fig. 4).

The main contribution in this aspect will be the addition of the semantically enhanced fourth level of the SCOR process model from each company to the mapping process, the hierarchical structure of the SCOR model should prove to be useful to enhance the results of this procedure. Once the schema information has been acquired and expressed in a unified model, further analysis and/or processing can be performed to identify a set of mappings between semantic entities being used in different business schemata. The goal is to provide such sets of mappings as input to translator tools to achieve interoperability between dispersed systems without modifying the involved schemata. The neutral representation of incoming schemata provides the basis for the identification of the relevant semantic entities being the basis of the mapping process; based on the annotation made with respect to the ontologies and on the logic relations identified between these ontologies, reasoning can identify correspondences on the semantic entity level and support

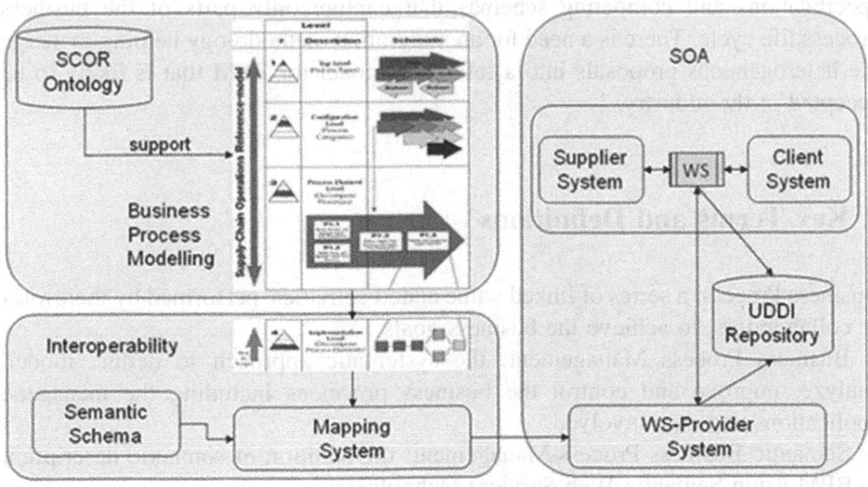

Fig. 4 Framework design

the mapping process. The definition of the mappings is done through the acquisition of the crucial features in the schemata of the source and target, giving them a conceptual representation, enriching this representation with semantic annotations and then using the system functionalities to synthesize and refine the mappings.

As result from the mapping system you get a transformation language (could be expressed in XSLT), it will be used to generate the web service which will be used as translator for the different schemas in the current information interchange. This way seamless interoperation at the process level is achieved including these translators in the executable process model of the workflow.

4 Conclusions and Outlook

An executable business process model has to specify not only the control flow, but also data flow, interfaces to the services that implement the activities, and run-time binding to service implementations. In traditional Business Process Management (BPM) many of these specification steps cannot be handled automatically or facilitated by appropriate tools, since the meaning of process artifacts is not captured in the process model. Furthermore, design artifacts such as existing process models or existing service interfaces are hardly reused in practice. The Semantic BPM (SBPM) approach addresses these challenges by annotating semantic descriptions to process artifacts, yielding the opportunity to ease a business analyst's job by providing tools that facilitate various parts of the BPM lifecycle. Future solutions should address the usage of the new semantics-enabled techniques that can be used when creating an executable process model.

The lack of a commonly accepted schema is still a major handicap for business process management. Competing standardization bodies have proposed numerous specifications and competing schemas that capture only parts of the business process life cycle. There is a need for an integration methodology helping to merge the heterogeneous proposals into a reference model for BPM that is likely to be accepted in the industry.

5 Key Terms and Definitions

Business Process: a series of linked value added activities, performed by their roles or collaborators, to achieve the business goals.

Business Process Management: the systematic approach to define, model, analyze, improve and control the business processes including the managers, applications and data involved.

Semantic Business Process Management: the addition of semantic description to BPM using Semantic Web Services technologies.

Ontology: a rigorous and exhaustive organization of some knowledge domain that is usually hierarchical and contains all the relevant entities and their relations.

Service Oriented Architecture: a flexible set of design principles used during the phases of systems development and integration. SOA provides a loosely-coupled suite of services that can be used within multiple business domains.

Business Process Integration: refers to the capability to orchestrate business processes in spite of its modeling language, securing interoperability of the data involved in the messages exchanged.

References

1. Abramowicz, W., Filipowska, A., Kaczmarek, M., Kaczmarek, T.: Semantically enhanced Business Process Modelling Notation. In: Proceedings of the Workshop on Semantic Business Process and Product Lifecycle Management (SBPM), Innsbruck, Austria (2007)
2. Andrews, T., Curbera, F., Dholakia, H., Goland, Y., Klein, J., Leymann, F., Liu, K., Roller, D., Smith, D., Thatte, S.: Business Process Execution Language for Web Services, Version 1.1. Specification, BEA Systems, IBM Corp., Microsoft Corp., SAP AG, Siebel Systems (2003)
3. Aslam, M., Auer, S., Shen, J.: From BPEL4WS Process Model to Full OWL-S Ontology. In: The 3rd European Semantic Web Conference, Budva, Montenegro (2006)
4. Business Process Modeling Notation Specification. OMG Final Adopted Specification, Jan 2009 (2009)
5. Dayal, U., Hsu, M., and Ladin, R.: Business Process Coordination: State of the Art, Trends, and Open Issues. In: Proceedings of the 27th international Conference on Very Large Data Bases (2001)
6. De Nicola, A., Mascio, T.D., Lezoche, M., Tagliano, F.: Semantic Lifting of Business Process Models. In: Proceedings of the 12th Enterprise Distributed Object Computing Conference Workshops EDOCW 2008 (2008)
7. Fantini, P., Savoldelli, A., Milanesi, M., Carizzoni, G., Koehler, J., Stein, S., Angeli, R., Hepp, M., Roman, D., Brelage, C., Born, M.: SUPER Deliverable D2.2: Semantic Business Process Life Cycle. http://www.ipsuper.org/res/Deliverables/M12/D2.2.pdf Accessed 2009
8. Hepp, M., Roman, D.: An Ontology Framework for Semantic Business Process Management. In: Proceedings of Wirtschaftsinformatik 2007, Karlsruhe, Germany, pp. 1–18 (2007)
9. Hepp, M., Leymann, F., Domingue, J., Wahler, A., Fensel, D.: Semantic Business Process Management: A Vision Towards Using Semantic Web Services for Business Process Management. In: Proceedings of the IEEE—ICEBE (2005)
10. Karagiannis, D.: BPMS: Business process management systems. ACM SIGOIS Bull. **16**(1), 10–13 (1995)
11. Ko, R.K.L., Lee, S.S.G., Lee, E.W.: Business process management (BPM) standards: a survey. Bus. Process Manage. J. **15**(5), 744–791 (2009)
12. Koskela, M., Haajanen, J.: Business Process Modeling and Execution: Tools and Technologies. Report for the SOAMeS Project. VTT Research Notes 2407, VTT Technical Research Centre of Finland (2007)
13. Kotinurmi, P., Haller, A., Oren, E.: Ontologically Enhanced RosettaNet B2B Integration. In: Garcia, R., (ed.) Semantic Web for Business: Cases and Applications, pp. 194–221. IGI Global (2009)
14. Lautenbacher, F., Bauer, B.: A Survey on Workflow Annotation and Composition Approaches. In: Proceedings of the Workshop on Semantic Business Process and Product

Lifecycle Management (SBPM) in the context of the European Semantic Web Conference (ESWC 2007), Innsbruck, Austria (2007)
15. Mendling, J., Neumann, G.: A Comparison of XML Interchange Formats for Business Process Modelling. Workflow Handbook 2005, Future Strategies Inc, Florida , USA (2005)
16. Ouyang C., van der Aalst, W., Dumas, M., ter Hofstede, A.: Translating BPMN to BPEL. BPM Center Report BPMcenter.org, available at http://is.tm.tue.nl/staff/wvdaalst/BPMcenter/reports.htm.Accessed 2006
17. Preist, C., Cuadrado, J. E., Battle, S., Williams, S., Grimm, S.: Automated Business-to-Business Integration of a Logistics Supply Chain using Semantic Web Services Technology. In: Proceedings of 4th International Semantic Web Conference, pp. 987–1001. Springer, (2005)
18. Recker, J.C., Mendling, J.: On the Translation between BPMN and BPEL: Conceptual Mismatch between Process Modeling Languages. In Latour, T., Petit, M. (eds.) 18th International Conference on Advanced Information Systems Engineering (2006)
19. Recker, J., Indulska, M., Rosemann, M., Green, P.: An Exploratory Study of Process Modeling Practice with BPMN. BPMCenter Report, No. BPM-08-12, www.BPMCenter.org. Accessed 2008
20. Shapiro, R.: A Comparison of XPDL, BPML, and BPEL4WS. OASIS XML Cover Pages (2002)
21. Simchi-Levi, D., Kaminsky, P., Simchi-Levi, E.: Designing and Managing the Supply Chain: Concepts, Strategies, and Case Studies, McGraw-Hill, New York (2000)
22. van der Aalst, W.M.P., ter Hofstede, A.H.M., Weske, M.: Business Process Management: A Survey. In: Proceedings of the Business Process Management: International Conference, BPM 2003, Eindhoven, the Netherlands, 26–27 Jun 2003
23. Weber, I., Hoffmann, J., Mendling, J., Nitzsche, J.: Towards a Methodology for Semantic Business Process Modeling and Configuration. In: Service-Oriented Computing—ICSOC Workshops: ICSOC 2007, international Workshops, Vienna, Austria (2007)
24. White, S.: Process Modeling Notations and Workflow Patterns. Workflow Handbook 2004, pp. 265–294. Future Strategies Inc, Florida, USA (2004)

Additional Reading Section

25. e3-value, http://www.e3value.com
26. FUSION Project, http://www.fusionweb.org
27. Ghalimi, I.,McGoveran, D.: Standards and BPM. BPM.COM (2005)
28. Havey, M.: Essential Business Process Modeling, O'Reilly Media, Inc., Sebastopol, CA 95472, USA (2005)
29. Hepp, M., Hinkelmann, K., Karagiannis, D., Klein, R., Stojanovic, N.: In: Proceedings of the Workshop on Semantic Business Process and Product Lifecycle Management (SBPM 2007) in conjunction with the 4th European Semantic Web Conference (ESWC 2007). http://sbpm2007.fzi.de. (2007)
30. Hepp, M., Hinkelmann, K., Karagiannis, D., Klein, R., Stojanovic, N.: In: Proceedings of the 4th International Workshop on Semantic Business Process Management (SBPM2009) collocated with ESWC 2009. http://sbpm2009.fzi.de. (2009)
31. Hill, J.B., Pezzini, M., Natis, Y.V.: Findings: Confusion Remains Regarding BPM Terminologies. Gartner Research, Vol. ID Number: G00155817. (2008)
32. Karagiannis, D., Woitsch, R., Utz, W., Eichner H., Hrgovcic, V.: The Business Process as an Integration Platform: Case Studies from the EU-Projects LD-Cast and BREIN. In: Proceedings of the 3rd Workshop on Semantic Business Process Management (SBPM 2008) in conjunction with the 5th ESWC 2008 (2008)
33. Khan, R.M.: What Standards Really Matter for BPM. In: Business Process Trends (2005)

34. Recker, J.C.: A Study on the Decision to Continue Using a Modeling Grammar. In: 13th Americas Conference on Information Systems, Keystone, Colorado (2007)
35. Smith, H.: Enough is enough in the field of BPM: We don't need BPELJ: BPML semantics are just fine. BPM3 (2004). http://www.fairdene.com/bpelj/BPELJ-Enough-Is-Enough.pdf Accessed (2013)
36. SUPER Project, http://www.ip-super.org
37. van der Aalst, W.M.P.: Don't go with the flow: Web services composition standards exposed. IEEE Intell. Syst. **18**(1), 72–76 (2003)
38. van der Aalst, W.M.P.: Patterns and XPDL: A Critical Evaluation of the XML Process Definition Language. In: QUT Technical report, FIT-TR-2003-06, Queensland University of Technology, Brisbane, 2003 (2003)
39. van der Aalst, W.M.P.: Business Process Management: A Personal View. Bus. Process Manage. J. **10**(2), 5 (2004)
40. Wohed, P., van der Aalst, W.M.P., Dumas, M., ter Hofstede, A.H.M.: Analysis of Web Services Composition Languages: The Case of BPEL4WS. Conceptual Modeling-Er 2003. In: Proceedings 22nd International Conference on Conceptual Modeling, Chicago, IL, USA (2003)
41. Wohed, P., van der Aalst, W.M.P., Dumas, M., ter Hofstede, A.H.M.: Pattern-Based Analysis of BPEL4WS. Department of Computer and Systems Sciences, Stockholm University/The Royal Institute of Technology, Sweden (2003)
42. Wohed, P., van der Aalst, W.M.P., Dumas, M., ter Hofstede, A.H.M., Russell, N.: On the Suitability of BPMN for Business Process Modelling. Business Process Management, pp. 161–176. Springer, Berlin (2006)
43. Woodley, T., Gagnon, S.: BPM and SOA: Synergies and challenges. WISE, pp. 679–688. Springer, Heidelberg (2005)
44. zur Muehlen, M.: Tutorial—Business Process Management Standards. In: 5th International Conference on Business Process Management (BPM 2007), Brisbane, Australia, pp. 13 (2007)

Chapter 6
Integrating ERP with Negotiation Tools in Supply Chain

Paolo Renna

Abstract Today marketplaces are characterized by a high degree of dynamism where companies need to operate by matching agility and efficiency. Supply Chain Management has become the crucial key to competitive advantage and traditional tools for enterprise management, such as ERP or MRP II, need to gain capabilities to allow inter-organizations relationships. The research presented in here aims at developing an Extended Agent Based ERP for enterprises relationship coordination. In particular, we refer to a vertical supply chain where suppliers operate under Assembly to Order (ATO) modalities; furthermore, suppliers are organised in a "buyer centric e-marketplace" by the assembler. In such a context, through the use of Multi Agent System (MAS) technology, the coordination of the ERP systems is approached by considering the development of an electronic negotiation system. The simulations conducted highlight the functionality of the cooperation among the ERP systems and the improvement of the performance both for customers and suppliers.

Keywords Supply · Chain · Enterprise resource planning · Multi agent systems · Cluster · E-procurement · Discrete event simulation

1 Introduction and Motivation

Today marketplaces are characterized by a high degree of dynamism due to an environment in which end users are more and more involved in product customization. Moreover, market globalization requires companies to operate in a wide and complex international market by matching agility and efficiency. Nowadays, even small and medium-sized enterprises increasingly rely on international

---

P. Renna (✉)
School of Engineering, University of Basilicata, Via dell'Ateneo Lucano, 10, 85100 Potenza, Italy
e-mail: paolo.renna@unibas.it

M. Mora et al. (eds.), *Engineering and Management of IT-based Service Systems*, Intelligent Systems Reference Library 55, DOI: 10.1007/978-3-642-39928-2_6,

networks of suppliers, distributors and customers to improve their global competitiveness [1]. Electronic Business (EB) can be defined as the "*use of information technology for improving business relations between trading partners*". EB aims at reinventing the business process, by reducing the cost, enhancing customer satisfaction and finally increasing enterprises profits by integrating the individual information technology for automatic exchange of business related information. The basic expectation of adopting EB approaches is to improve optimal resource allocation at the Supply Chain (SC) level in order to improve SC surplus. This requires not only intra-enterprise integration, but also inter-enterprise integration [2]. In such situation, Supply Chain Management (SCM) has become the crucial key to gain competitive advantage. After 2 decades of streamlining internal operations, boosting plant productivity, improving product quality and reducing manufacturing costs, companies are focusing on SC strategies as the next frontier in organizational excellence. In this network economy, markets are becoming more transparent, customer demands are being met in a more customized manner [3] and, in general, the rate of change in the business world keeps increasing [4]. Interestingly enough is that, at the same time, the Enterprise Resource Planning (ERP) is sweeping across industry. ERP, the logical extension of the Material Requirements Planning (MRP) systems of the 1970s and of the Manufacturing Resource Planning (MRP II) systems of the 1980s, is now a standard in industry. It is a comprehensive transaction management system that integrates many kinds of information into a single database. Researchers have pointed to information system fragmentation as the primary culprit for information delays and distortions along the Supply Chain [5]: an ERP system could potentially enhance transparency across the SC by eliminating information distortions and increasing information velocity by reducing information delays. Hence, there are reasons to believe that ERP adoption could be associated with significant gains in SC effectiveness. However, at SC level ERP systems need to be integrated in order to work in a synchronized way as a whole Extended SC ERP system. This is the key of success of SCM, to co-ordinate information among SC actors in order to improve resource effectiveness and efficiency along the whole SC. Indeed, sharing data regarding performances such as lead times, quality specifications, return status, etc., helps SC partners to identify and overcome the bottlenecks in their development [6]. This chapter proposes an approach to integrate ERP systems at SC level by using MAS technology [7] and negotiation [8]. This approach does not require integration at single ERP level, because information exchange, processing and transaction are guaranteed by Agents interfacing with each ERP system. In this way, each SC actor maintains its own information system and SC synchronization is obtained through the co-ordination of distributed decision makers. This research refers to a vertical integrated SC where an assembler has organised its own suppliers in a "buyer centric e-marketplace" and a test platform is developed.

In particular, each suppliers and customer ERP system has been designed and developed by using an Access® database while, agents have been developed by using JAVA® open source platforms. After a detailed analysis of the developed system, it has been tested the effectiveness and the efficiency of the proposed approach. In particular, it has been developed two cases studies: the first represents the situation in which each supplier negotiates the terms of transaction directly with the assembler; the second one, instead, explores the situation in which the suppliers are organised in stable clusters

The chapter is structured as it follows: Sect. 2 provides a literature review, while in Sect. 3 the main findings of the research are discussed. The context of the proposed approach is described in Sect. 4; in Sect. 5 the agent based e-marketplace model is presented; in Sect. 6, the local ERP is briefly described; the cluster approach proposed is presented in Sect. 7; in Sect. 8 is described the simulation environment developed while in Sect. 9 the case study proposed. Finally, simulation results are discussed in Sect. 10 and conclusions, and further research paths are withdrawn in Sect. 11.

2 Literature Review

Many researchers have been proposed framework on procurement approaches among customers and suppliers. Few researchers have been developed to integrate several tools together as ERP, negotiation and suppliers cluster.

Cantamessa et al. [9] presented a conceptualization of a Multi Agent architecture to support extended enterprises. The research proposed an innovative architecture for distributed process and production planning in a manufacturing environment. The architecture is based on the agent technology and it aims at automating most of the phases within the supply chain of a manufacturing network. The architecture is the result of a research project whose ultimate goal is to build a business to business application for distributed process and production planning in manufacturing networks of Small and Medium Enterprises

Ash and Burn [10] reviewed the results of a 3 year study into Internet enabled ERP implementations around the world. The study identified different stages of growth with differing sets of problems at each stage. A framework for e-business change was used to evaluate the mature stage of e-ERP in six international organisations. The emergent model proposes various antecedents to successful e-business change management in ERP environments. A case study of the first B2B e-business integration with Dell Computer Corporation and its largest corporate customer is examined in the context of this model. The case demonstrates the integration of ERP and non-ERP systems, using Web-based technologies, to optimize an overall B2B value chain. Finally, the paper emphasised the role of change management and cultural readiness when adopting e-business solutions and identifies critical areas for future research.

Rudberg and Olhager [11] analyzed the manufacturing networks and supply chains from operation strategy prospective. They presented a typology for the analysis of network systems resulting in four basic network configurations. Therefore, four coordination approaches have been proposed for further research in the context of integrating manufacturing network and supply chain theory.

Cavalieri et al. [12] developed a Multi-Agent model to implement a vertical and lateral coordination among components of a non-centralized distribution chain. This approach has been implemented in Java language in order to allow for the development of a Web-based decision support system that can perform the coordination strategies simulated. The models have been applied to a real two-level distribution system of an electromechanical company, made up of a supplier and a geographically distributed network of retailers.

Kang and Han [13] developed a broker-based synchronous transaction algorithm that would guarantee a more fair and efficient transaction deal for both sellers and buyers. This algorithm, implemented by Visual C++, showed better performance in every aspect in the experiment for comparison with the current two model types. The number of transactions increased by 21 % and the prices were adjusted up to 280 % more efficiently in some transaction cases.

Argoneto et al. [14] presented an innovative approach for a private neutral linear e marketplace. This approach realizes a full integration between the customer order and the supplier planning activity as they would share the same Enterprise Resource Planning System. Actually, the customer and the supplier system interact through an agent-based network. Furthermore, a set of Value Added Services is proposed in the e-marketplace and specifically negotiation and coalition support services. Finally, a proper test environment has been designed and modeled in order to measure "the stay-together economy" achievable within the proposed innovative e-marketplace.

Argoneto et al. [15] proposed a Multi Agent Architecture to integrate and coordinate different Enterprises by negotiation methodologies. The research presented in this chapter is an extension of this paper.

Perrone et al. [16] developed an innovative tool for manufacturing enterprise networks organized into an e-marketplace. They have been conceptualized, designed, implemented and tested a Multi Agent Architecture. The topics developed concern: the development of technological planning; multi attribute negotiation mechanism; production planning as a support to negotiation mechanism; the development of a simulation environment by open source package; the distributed architecture is benchmarked with a centralized approach.

Lea et al. [17] proposed a prototype multi-agent enterprise resource planning (MAERP) system that utilizes the characteristics and capabilities of software agents to achieve enterprise wide integration. A software agent is a self-contained, autonomous software module that performs assigned tasks from the human user and interacts/communicates with other applications and other software agents in different platforms to complete the tasks. Four types of intelligent software agents (coordinating agents, task agents, data collecting agents, and user interface agents) are examined and discussed in the proposed MAERPS architecture. They demonstrated how the proposed prototype MAERP system takes advantage of existing information systems among various functional areas to achieve the system integration of commercially available ERP systems, while avoiding numerous problems encountered during a typical ERP implementation.

Xu and Beamon [18] developed a framework that enables organizations to select appropriate coordination mechanisms, based on relative costs and the characteristics of their specific operating environment.

Gunasekaran et al. [19] analyzed some case experiences on Agile Manufacturing and Supply Chain Management and developed an integrated framework for Responsive Supply Chain. The proposed framework can be employed as a competitive strategy in a networked economy in which customized products/services are produced with virtual organizations and exchanged using e-commerce.

Renna [20], Renna and Argoneto [21] proposed negotiation policies, customer's tactics and coalition tools as a value added services in e-marketplaces. The e-marketplace scenario is a private neutral linear owned by a third part with catalogue based procurement actions. A Multi Agent System methodology is used to implement the architecture of the e-marketplace.

Saeed et al. [22] examined the inter-organizational system (IOS) configuration choices made by firms with different supply chain integration profiles. Their results support the notion that successful firms sequence the configuration of IOS characteristics toward effectively developing and supporting their supply chain process capabilities.

Olson and Staley [23] studied the open-source software (OSS) based ERP system for small business. The paper reported a case study based on ERPlite and xTuple software. While both systems worked from a technical perspective, both failed due to economic factors. While these economic conditions led to imperfect results, the case demonstrates the feasibility of OSS ERP for small businesses. Both experiences are evaluated in terms of risk dimension.

Shafiei et al. [24] investigated both the fields of ERP and decision support systems (DSS) in a multi-enterprise collaboration environment.

They combined their own insight with respect to multiple-enterprise collaboration via the integration of ERP and DSS to propose a set of high-level and medium-level system frameworks.

Nazemi et al. [25] illustrated a complete literature survey on ERP systems. Among the several findings of the paper, they argued as the major case studies analyzed the SAP systems not suitable for small and medium enterprises.

Chen [26] investigated the coordination mechanism for supply chain with one manufacturer and multiple competing suppliers in the electronic market. They studied two conventional price-only policies, including wholesale price policy and catalog policy, based on the reverse Vickrey auction.

3 Main Findings

The main limits of the literature can be summarized as follows.

Most of the researches investigated only one area of the network; for example, on the negotiation problems, selection of the suppliers, coalition approaches, and so on. Some research presented the framework that supports the proposed approach, but a numerical analysis is not performed in order to compare the performance measures. This lack of the literature leads to reduce the evidence of the real advantages that the small and medium enterprises can gain by these tools.

The research proposed in this chapter overcomes these limits combining the development of a framework to support the proposed approach and the simulation to evaluate the benefits in a quantitative manner.

The main issues addressed by the chapter are the following. The fist step is the development of a Multi Agent Architecture as a tool to integrate the procurement process in a supply chain. The design process is performed using workflow management methodologies integrating UML and IDEF tools. A supplier cluster approach has been proposed as a coalition mechanism among the suppliers. Furthermore, the author proposes the use of discrete-event simulation to test the proposed approach and to evaluate the economic value of adopting the proposed planning and negotiation tools in e-marketplaces.

The simulation environment and agent architecture is based on open source IT tools for the software platform development.

The motivations of the methodologies used to support the proposed research are the following. Multi agent system seems to be the most promising technology to be used in building these systems [7].

Workflow analysis tools seem to be the most promising methodology to engineering e-business Vale Added Services (VAS) design [27]. The open environment such as the Internet and the WWW, and open IT tools, such as Java and XLM, are considered the reference IT technologies for developing these systems. These IT technologies support the development of the WEB platform in which the proposed approach will be implemented. The proposed approach uses workflow management methodologies for the design activities, agent-based technologies for the implementation phases, and open source IT tools for the software platform development. Furthermore, the author proposes the use of discrete-event simulation to test the proposed approach and to evaluate the economic value of adopting the proposed planning and negotiation tools in e-marketplaces.

4 The Context

The reference context is an Assembly To Order environment, where only systems elements, modules or subassemblies are in stock at the manufacturing center, whereas the final assembly takes place based on a specific customer order; some examples can be: cars, computers, etc.

In this chapter, the products are different bicycle typology, in particular, three-part mix. The assembler manages a supplier database, in this database, there are suppliers registered and qualified that can supply all the items requested by the assembler.

Each supplier is characterized by a manufacturing system of five workstations machine in order to imply the lead time between the order and the delivery of the items.

The procurement process among suppliers and assembler is performed by a multi-attribute negotiation mechanism that is activated for each order requested to the assembler by the buyer.

The following assumptions are considered:

- the supply chain is constituted by one level;
- the distances among suppliers and assembler are not implied;
- delivery lead times are neglected.

Figure 1 shows the e-marketplace structure through the interaction with external actors. As the reader can notice, three kinds of actors have been located:

- the buyer, is the generic client, who is allowed to input an order of final product and sets the order specifications;
- the customer, is the generic registered e-marketplace client, who is allowed to provide its Bill of Material (BOM) explosion, production planning information by its ERP and negotiation strategies;

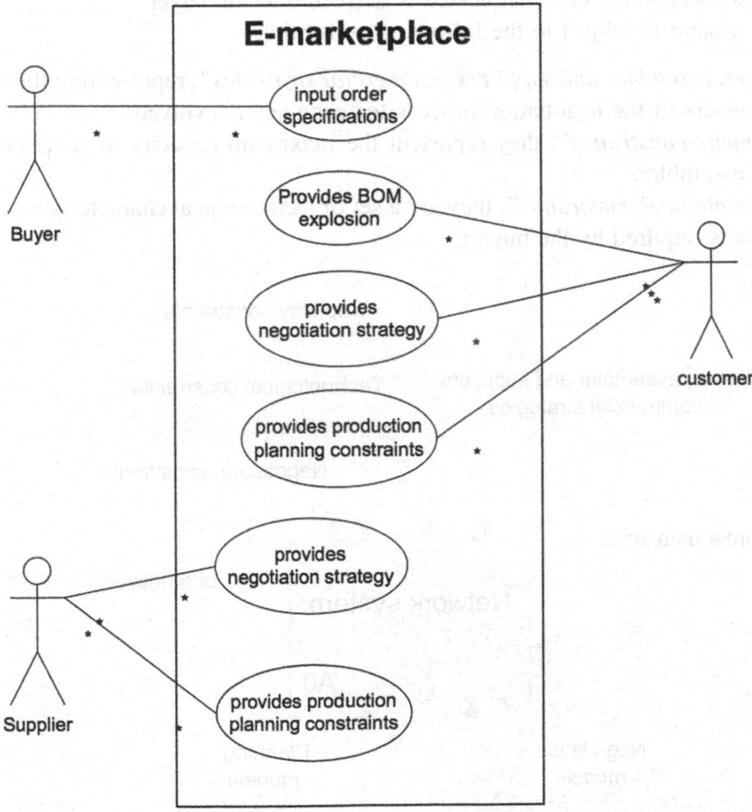

Fig. 1 Use case diagram

- the supplier, is the generic registered e-marketplace supplier, who provides negotiation strategies and planning information.

Once the buyer inputs its order, the procurement transaction proceeds automatically being handled respectively by the customer and supplier agents.

The architecture that constitutes the framework of the network is formalized through the use of the IDEF0 formalism [28].

Every Diagram (or Node) is defined through some boxes, which represent functions or activities performed by the system. The IDEF0 process starts from the identification of the top level diagram that can be decomposed in several decomposition diagrams ("child" in IDEF0 terminology).

The context in which the system operates has been defined through the top level diagram of Fig. 2 that distinguishes the system from the external environment.

The global input of the system is given by the "*order data input*": it deals with the activity of the buyer who requests an order. The order is characterized by the commercial data (volume, price and date of delivery) and technological information (product typology).

The output of the system is the "*order delivery*", if the negotiation process achieves success, the order requested is delivered to the buyer.

The system is subject to the following constraints:

- "*buyer, assembler and suppliers commercial strategies*", representing the initial parameters of the negotiation process for each actor involved;
- "*capacity constraints*", they represent the maximum capacity of suppliers and final assembler;
- "*technological constraints*", they are a set of technological characteristics of the products required by the buyer;

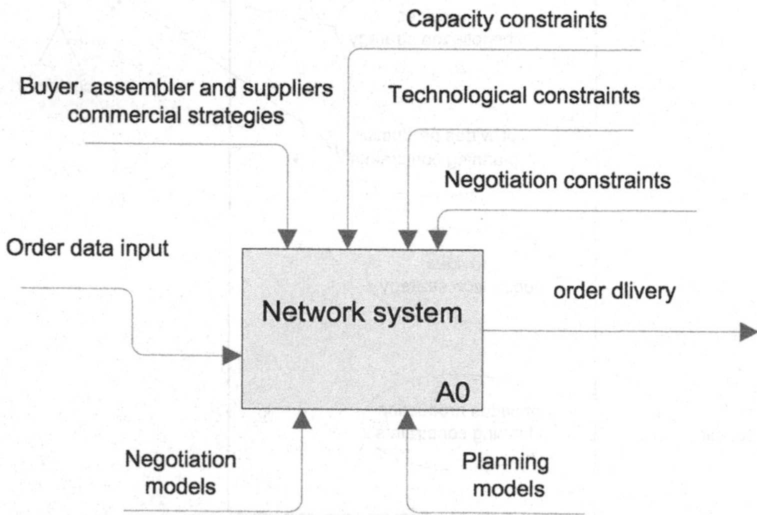

Fig. 2 Node A-0 context diagram

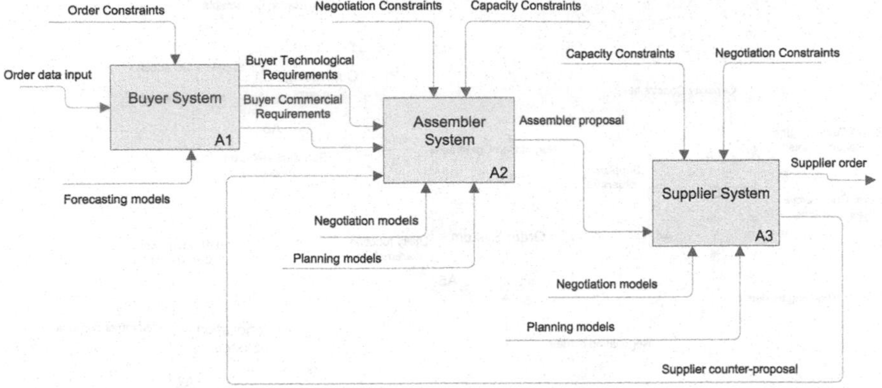

Fig. 3 Node A-0—level 1—overall System

- "*negotiation constraints*", that is the rules for information sharing and exchange among buyer, assembler and suppliers.

The system operates through the following mechanisms:

- "*negotiation models*", they provide the operational scheme for the bargaining among the agents that negotiate;
- "*planning models*", they provide the operative scheme for production planning activity to support the assembler and the suppliers.

The Network system can be decomposed in a structure composed of three parts (see Fig. 3): *Buyer system*, *Assembler system* and *Supplier system*.

The input of the buyer system is the global input of the system ("*order data input*"). The Buyer outputs are the "*Buyer technological requirements*" of the order (product typology) and the "*buyer commercial requirements*" (volume, price and due date). These outputs are computed on the basis of market environment ("*Forecasting models*" and "*Order constraints*").

The two outputs of the Buyer system constitute the inputs of the Assembler system that processes these data to produce its "*Assembler proposal*" (volume, dude date and price of raw products). The "*assembler proposal*" is computed on the information provided by the "*planning models*", the "*negotiation models*", "*negotiation constraints*" and the "*capacity constraints*".

The "*assembler proposal*" is the input of the supplier system. The outputs of the supplier system are the: "*supplier order*" and the "*Supplier counter-proposal*". The "*Supplier counter-proposal*" is elaborated on the basis of the information provided by the "*planning models*", "*negotiation models*", "*negotiation constraints*" and "*capacity constraints*".

The Assembler system is composed by four sub-system showed in Fig. 4: the *ERP system*, *Order system*, *Contacting system* and *Negotiation system*.

The ERP system starts when receives the buyer technological and commercial requirements, it performs the planning models and subject to the capacity

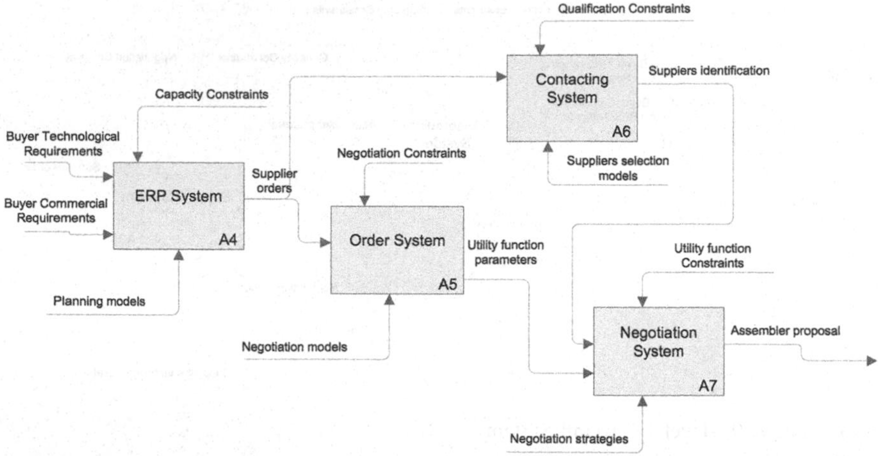

Fig. 4 Node A-2—level 2—assembler System

constraints computes the "*Supplier orders*". The "*supplier orders*" information is transmitted to the order system and contacting system.

The order system provides to the negotiation system the "*utility function parameters*" for the negotiation process with the suppliers. The "*utility function parameters*" are computed by the "*negotiation models*" and it is subject to the "*negotiation constraints*". The contacting system utilizes the "*supplier orders*" and the "*suppliers selection models*" in order to select the suppliers. The suppliers selected are subject to the "*qualification constraints*" and their identifications are transmitted to the negotiation system.

The negotiation system formulates the "*assembler proposal*" through the "*utility function parameters*" and the "*suppliers identification*" and it is subject to "*utility function constraints*".

The *Supplier system*, shown in Fig. 5, consists of the following three subsystems: the *ERP system*, the P*roduction planning system* and the *Negotiation system.*

The ERP system computes the "*Production alternatives*" by using the "*Assembler proposal*" and "*Planning models*", it is subject to "*capacity constraints*". The "*Production alternatives*" is the input of the Production planning system that finds out an optimal resource allocation plan ("*production orders*") for each production alternative; this allows the Production planning system to build a function mapping transmitted to the negotiation system. The negotiation system computes a counter-proposal through the "*negotiation strategies*" and "*production orders*" information, and it is subject to the "*production constraints*".

The combination of UML and IDEF modeling methodologies allow to support the IT-based system engineering projects. The use modeling approaches can achieve accurate requirements for the subsequent software design activity.

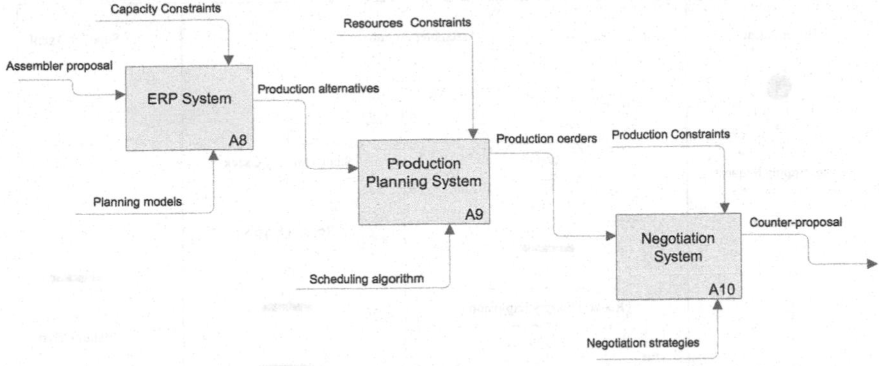

Fig. 5 Node A-3—level 2—supplier system

The UML provides a set of modeling notation based on object-oriented software that enables IT system to be reused, reconfigured and scaled-up and down. These enterprise modeling-methodologies define the requirements for the development of the collaborative software tool with the objectives of eliminating inconsistencies and providing the right information to stakeholders at the right time. The opportune development of the modeling contributes to fill the communication gap among the organizations and the system-manager; the diffusion of the manuals and procedures for the use of the platform among the participants.

5 The Agent-Based E-Marketplace Model

The Agent Based Model adopted in here is the following: a Buyer Agent (BA) submits, to the Assembler Agent (AA), an order request consisting of the array $\{V, DD, P\}$, being V the required Volume, DD the required Due Date and P the required price for the order. This last Agent explodes the BA order in its elementary components, $i \in \{1, 2, \ldots, I\}$, and, by running its own ERP system, it verifies the availability of an effective stock quantity, for each component, in its own magazine. The number of items, for each component i, needed to satisfy the BA demand, constitute the order to be submitted to all the Supplier Agents (SA) joining the e-marketplace. An Assembler ERP Agent retrieves this information from the Assembler ERP and puts it in the Supplier Networks as depicted in Fig. 6.

In particular, the activities showed in UML Activity Diagram (see Fig. 6) are the following:

- The process starts with the request of product submitted by the buyer agent to the customer agent; then, the Buyer Agent waits for the terms of commitment.
- The Supplier Agent receives the request by the Buyer Agent and applies the ERP software to explode the final product in items and formulate the request of its to the suppliers;

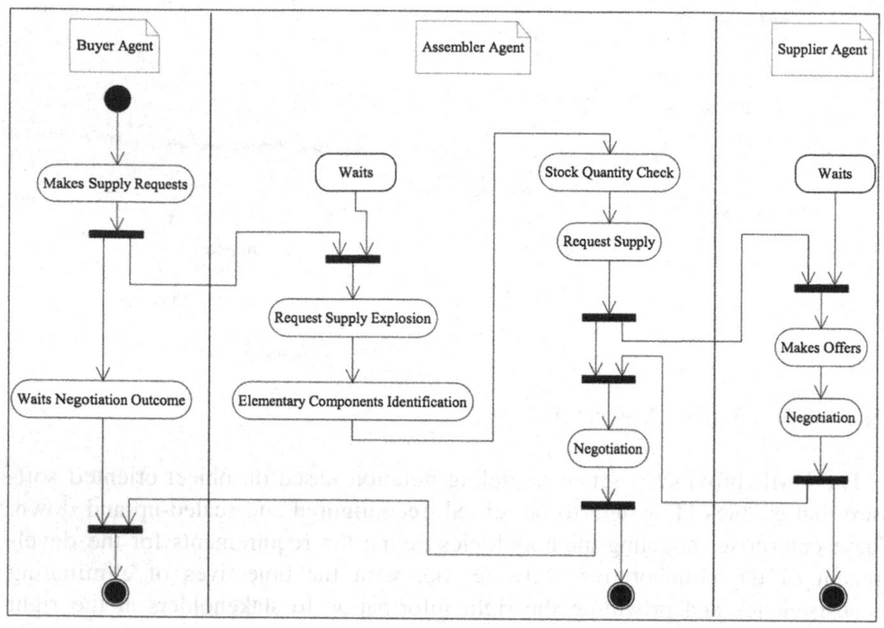

Fig. 6 Networked based model

- Then, the Assembler Agent computes the proposal and negotiates with the suppliers;
- The supplier Agent receives the proposal by the supplier Agent and computes the counter-proposal by the information of its ERP software.

5.1 Assembler agent model

The Assembler Agent is composed by the following agents:

ERP Agent: This agent is in charge with:

- exploding the billing of material of the BA order in its own components, $i \in \{1, 2, \ldots, I\}$;
- verifying the availability of the necessary items in the supplier magazine;
- sending to the Order Agent the component orders, that is, for each item *i*: $\{V_1, \ldots, V_i, \ldots, V_I\}$.

Order Agent defines the assembler utility function *uf* and its threshold level, *Z*, as it follows:

$$uf = \alpha\% + \beta\% + \gamma\% \tag{1}$$

where:

$$\alpha_i = \frac{|V_i - V_i^j|}{V_i} \tag{2}$$

$$\beta_i = \frac{DD_i^j - DD_i}{D_i} \tag{3}$$

$$\gamma_i = \frac{P_i^j - P_i}{P_i} \tag{4}$$

Being V_i, DD_i and P_i respectively the i-th component volume, due date and price as resulting from the bill of material explosion and V_i^j, DD_i^j P_i^j respectively the volume, the due date and the price offered by the jth supplier, $j \in \{1, 2, \ldots, J\}$, during the negotiation process supplier and $DD_i = DD - ALT$, where ALT is the Assembler Lead Time.

The expressions (2, 3, 4) compute the percentage difference between the supplier proposal and the customer request.

Contacting Agent: it is in charge with the supplier selection for each item i; once a supplier is selected its code is sent to the Negotiation Agent. Supplier selection is made by using available information such as contractors, factory status, and performance evaluation.

Negotiation Agent: the Agent gets in contact with the selected suppliers by providing them with the item order request, i.e. (V_i, DD_i, P_i) for each i. Afterwards, it evaluates all the supplier counter proposal; if any counter proposals has an utility greater than the threshold value, i.e. $!\exists / uf^j > Z, \forall\, j$, it starts with the next bargaining round r as far as the maximum steps number is reached ($r \leq r_{\max}$); otherwise, an electronic contract is stipulate. Figure 3 shows the AA's agents workflow.

5.2 Supplier agent model

The generic supplier agent j architecture, whose workflow is depicted in Fig. 4, consists of a:

ERP Agent: after having updated the supplier data base with the assembler order, it build a set of alternative strategies for responding to the assembler order; these alternatives are provided to the Supplier Negotiation Agent in order to be used during the negotiation as depicted in Fig. 5;

Production Planning Agent: the agent plans the production activities as defined by the ERP Agent;

Negotiation Agent: the agent negotiates with the Assembler negotiation agent. The activities of the UML Activity Diagram (Fig. 7) are the following:

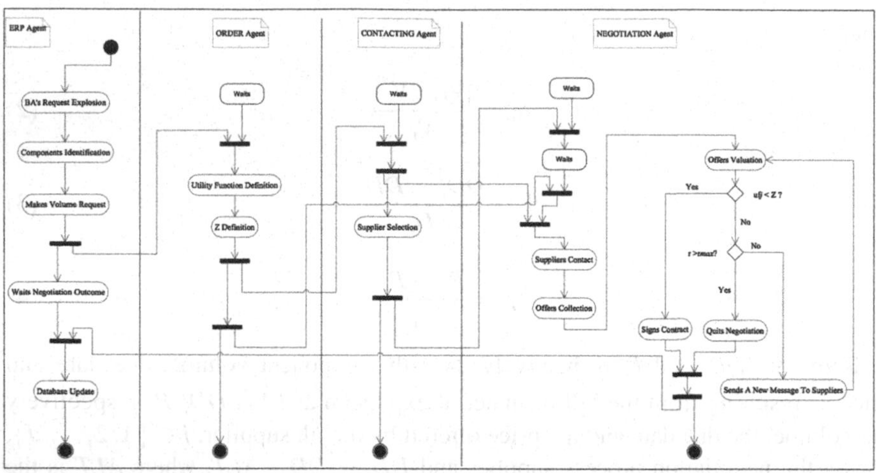

Fig. 7 The assembler agent model

- The ERP Agent explodes the final product requested by the buyer;
- The ERP Agent for each item identifies the item code and the ERP software releases the order item and manufacturing item.
- Then, The ERP Agent transmits the order information to the Order Agent and waits for the negotiation ending.
- The Order Agent computes the utility function and the threshold level for the negotiation process based on the information transmitted by the ERP Agent;
- The Contacting Agent selects the suppliers to contact for the negotiation.
- The Negotiation Agent contacts the Suppliers Agent and transmits to them the requested proposal;
- At each round of negotiation, the Negotiation Agent evaluates the counter-proposal submitted by the supplier agent. If the evaluation is positive signs the contract, otherwise it requests for a new counter-proposal. IF the maximum number of round is reached the negotiation end without agreement.

The activities of the UML Activity Diagram (Fig. 8) are the following:

- The ERP Agent performs the ERP software and builds the feasible production alternatives; it transmits this information to the negotiation agent and waits for the negotiation conclusion.
- The Negotiation Agent computes the counter-proposal and submits its to the Customer Agent. If an agreement is reached, the Negotiation Agent transmits the contract conditions to the ERP Agent and Production Planning Agent.
- The ERP Agent and Production Planning Agent update their database on the new contract signed.

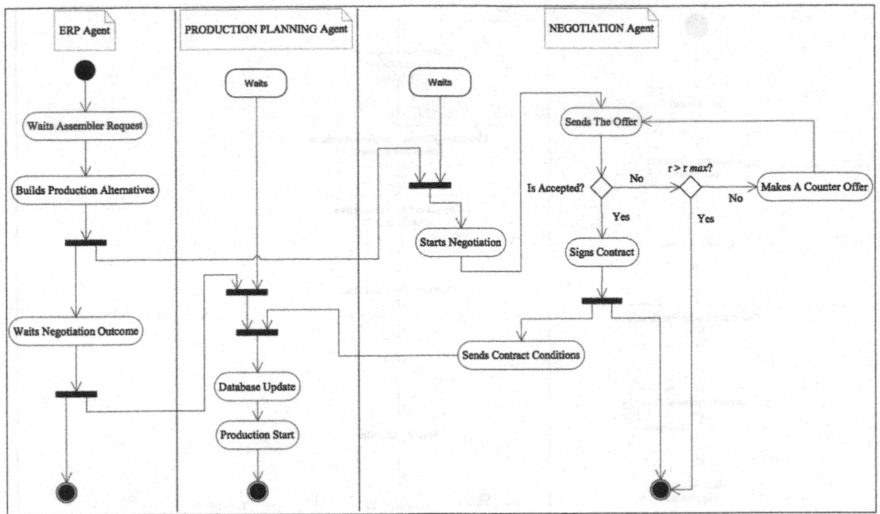

Fig. 8 Interactions between supplier agents

5.3 ERP Agent Proposal Algorithm

The activity "*Builds Production Alternatives*" works through the following steps (for each item *i*) (Fig. 9):

1. sets the counter proposal index $s = 1$ and $T = DD_i$;
2. the production capacity in *T* is evaluated by considering the resources workload needed to satisfy the assembler request;
3. the maximum volume deliverable in *T*, i.e. V_{max}, is computed considering the availability and the standard productivity (16 h/day) of the bottleneck resource;
4. the supplier's production costs (*C*) is calculated as $C = c \cdot V_{max}$, being *c* the unit component cost;
5. the price of the *i*-th item of the supplier *j* is computed by applying a mark-up strategy as in (5), where the mark-up coefficient *M* is computed as in (6):

$$P_i^{j(s)} = C \cdot (1 + M) \tag{5}$$

$$M = 0.25 - \left(0.15 \cdot \frac{T - DD_i}{n}\right) \forall T \tag{6}$$

As the reader can notice, *M* gets its maximum value, equal to 25 % of the item cost, when the order is dispatched within the request due date and the minimum value, equal to 10 % of *C*, when it is delivered within $T = DD_i + n$ periods;

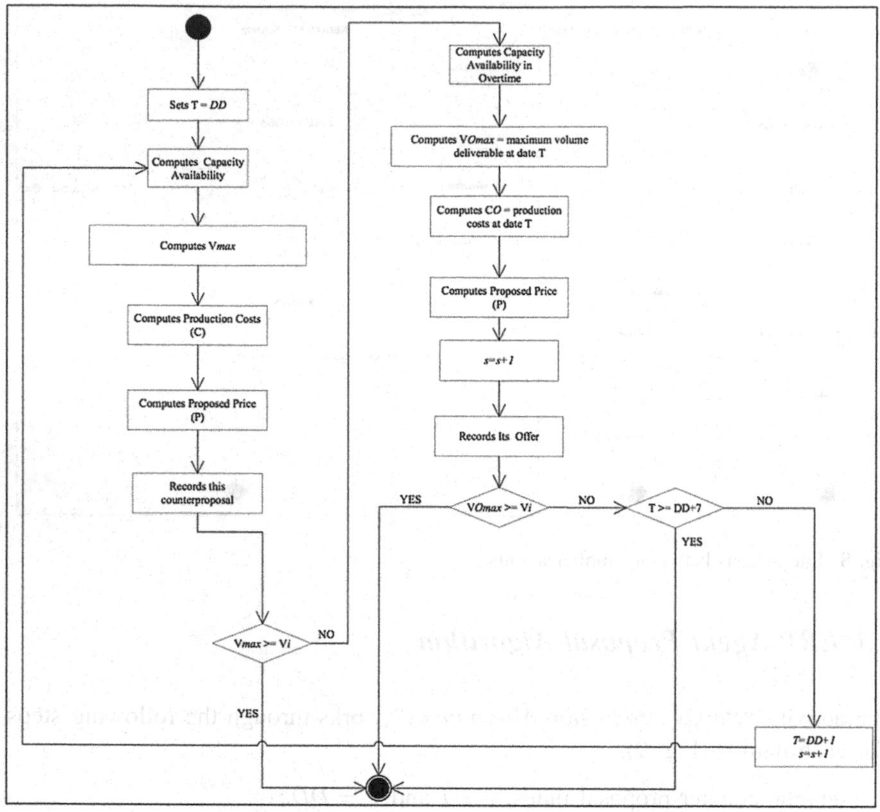

Fig. 9 Alternative proposals algorithm

6. sets $V_i^{j(s)} = V_{max}$ and $DD_i^{j(s)} = T$; the array $\left\{ V_i^{j(s)}, DD_i^{j(s)}, P_i^{j(s)} \right\}$ represents one possible counter-offer the supplier Negotiation Agent will negotiate with the Assembler;
7. if $V_{max} < V_i$, the supplier builds another counter-proposal by allocating the overtime production capacity available in T; let us indicate with VO_{max} the maximum volume that is possible to manufacture by considering an overtime production capacity of 8 h/day;
8. in this case the supplier's production costs (C) is calculated as $C = c_o \cdot VO_{max}$, being c_o the unit component cost in overtime capacity; also in this case the price is computed following the expressions (5) and (6);
9. sets $s = s + 1$, $V_i^{j(s)} = VO_{max}$, $DD_i^{j(s)} = T$ and the new value of $P_i^{j(s)}$; the array $\left\{ V_i^{j(s)}, DD_i^{j(s)}, P_i^{j(s)} \right\}$ represents another possible counter-offer the supplier Negotiation Agent will negotiate with the Assembler;

10. If still $VO_{max} < V_i$, the supplier increases the due date by setting $T = DD_i + 1$ and the algorithm starts from step 1; on the other hand, if $T > DD_i + n$ the algorithm ends.

5.4 Negotiation Protocol

The negotiation protocol is depicted in Fig. 10. The strategic behaviour pursued by the Supplier Agent looks for the maximization of its own utility, represented by the profit, which it gains when the bargaining reaches a positive end. It starts the negotiation by proposing the proposal, which would bring the highest profit, and, if it is not accepted, it reduces its gain ambition and selects, in the strategies set, the proposal with a lower profit value for the following round. The supplier a profit is computed by multiplying the price times the offered volume.

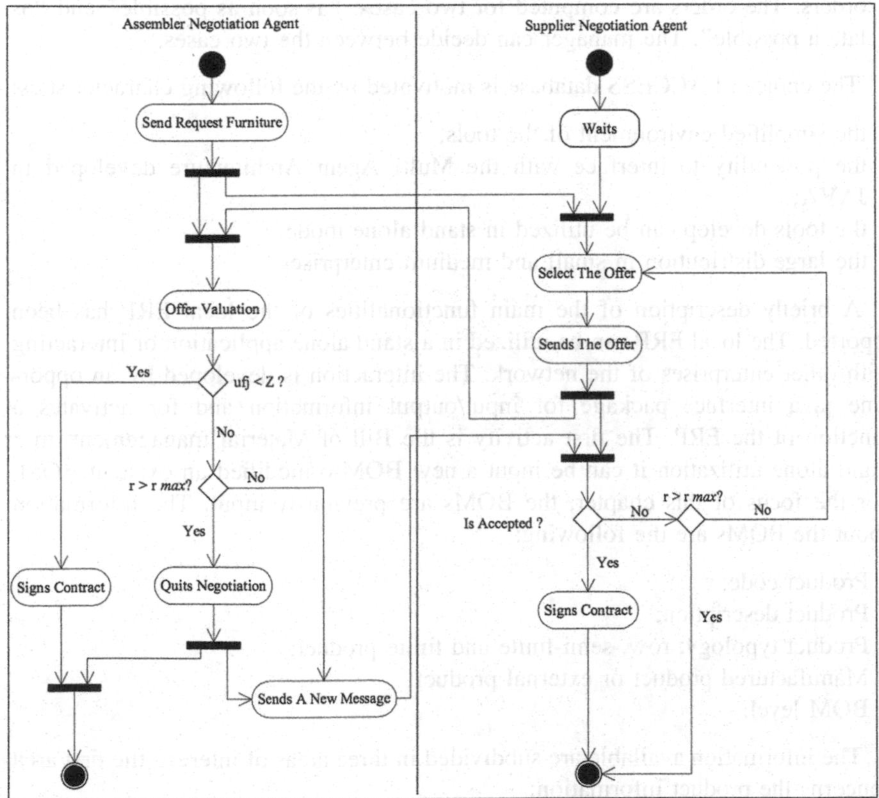

Fig. 10 The negotiation protocol

6 Enterprise ERP

In order to test the link between the negotiation mechanism and planning process an Enterprise Resource Planning is developed. The ERP developed is linked to each enterprise of the network, both customer and Suppliers. The ERP has been developed by the access database in order to plan the resources of the enterprise and provides information for the negotiation mechanism.

The features of the software are the following:

- Bills of Material; this section allows to insert/modify a BOM of the products and the attributes (quantity and lead time) for each item.
- Orders control; the section provides the information of all the orders. The orders released and the closed orders in order to evaluate the historical data of the suppliers.
- Inventory management, both raw and final products. The section allows to supervise the inventory level and provides the information to Manufacturing Resource Planning applying an inventory policy previously defined.
- Manufacturing Resources Planning. It generates the production and suppliers orders. The orders are computed for two cases: "as soon as possible" and "as late a possible". The manager can decide between the two cases.

The choice of ACCESS database is motivated by the following characteristics:

- the simplified environment of the tools;
- the possibility to interface with the Multi Agent Architecture developed in JAVA;
- the tools develop can be utilized in stand alone mode;
- the large distribution in small and medium enterprises.

A briefly description of the main functionalities of the local ERP has been reported. The local ERP can be utilized in a stand alone application or interacting with other enterprises of the network. The interaction is developed by an opportune java interface package for input/output information and for activates a function of the ERP. The first activity is the Bill of Material management, in a stand alone utilization it can be input a new BOM o modified an existent BOM. For the focus of this chapter, the BOMs are previously input. The information about the BOMs are the following:

- Product code;
- Product description;
- Product typology; row, semi-finite and finite product;
- Manufactured product or external product;
- BOM level.

The information available are subdivided in three areas of interest, the first area concerns the product information:

- Product suppliers, in case of external supply;

- Maximum price value; this is a maximum value of sale and it depends on opportune market analysis.
- Manufacturing or purchase costs;
- Manufacturing or purchase lot; it is the minimum lot size that can be manufacturing or supplied.
- Production lead time or supplier lead time;
- Standard productivity, it depends on the working time of the manufacturing resources.

The middle part of the mask concerns the "father" of the product under inspection.

In the bottom of the figure, there are the "son", those are the component necessary to complete the final product.

A customer can utilize the orders management function to insert an order submission, the fields are the following:

- The first field is the customer code number that identifies the customer. This code is assigned to customer in registration procedure.
- Date; this field contains the now date; therefore, it doesn't modify.
- A progressive number is assigned to the order submitted by the customer;
- A volume must be insert by the customer;
- A due date is indicated by the customer;
- Finally, the customer selects the code product requested.

The two functions that can be activated for the following operations are:

- A submission of a purchase order;
- An estimate purchase; in this case, the purchase order is suspended until the customer accepts or the validity period is end.

If the proposal is acceptable, the proposal is obtained by a planning on earliest and latest due date. If the proposal doesn't acceptable, that is the due date isn't match, then a proposal with earliest due date is proposal.

The production planning information of the ERP is shown in a mask. A the top of the mask the following information are available:

- Available of product in inventory;
- The request volume in order to satisfy the customer order;
- The above quantity is defined in terms of lot size;
- The due date of the product requested to the suppliers.

The ERP releases the production orders and the purchase orders. The production orders are planned on the available resources of the manufacturing system.

In case of purchase orders, the order data are transmitted to the supplier's network in order to start the negotiation process.

7 Cluster Case

In literature, several definitions are proposed to terms cluster. Porter [29] defined cluster as a group of firms engaged in similar or related activities within a national economy. Schmitz [30] on the other hand, defined cluster as a group of enterprises belonging to the same sector and operating in close proximity to each other. Porter [31] further defined clusters as geographically concentration of interconnected firms and institutions in a particular sector. The linkages existing between the firms are very important in strengthening competition.

Complimentarily is also used to explain the term cluster. "Clusters are sets of complementary firms (in production and service sectors) public, private and semi-public research and development institutions, which are interconnected by labour market and/or input–output and/or technological links" [32].

In this chapter, the suppliers are interconnected in horizontal integration in order to obtain a cluster. Therefore, a cluster of suppliers is able to provide all the items necessary to assembly the final product requested by the customer.

The main advantages of a cluster are the following:

- reduction of time to market; in case of development of new items, the collaboration among the enterprises of the cluster can reduce the development times.
- sharing risks; the negotiation with the customer is of the all cluster, then the disagreement costs are shared among the enterprise of the cluster.
- sharing knowledge; the collaboration makes easy the diffusion of knowledge.
- efficiency for common operations; the logistics service can be shared with high reduction of costs.

The main disadvantages of a cluster are the following:

- managerial difficult; the cluster has to be designed and provided by a managerial structure, for example, to resolve conflicts.
- relations among independent enterprises; the participants to the cluster are independent. Therefore, a trust problem can be occurred.

The life cycle of a cluster is characterized by the following main phases:

- selection of partners; in this phase, the future enterprises of the cluster have to be selected in order to pursue the objectives of the cluster.
- design and integration; IT technologies for communication and collaboration have to provided; rules and procedures have to de defined for the interactions among the enterprises.
- management; the management of the operational activities of the cluster.
- dissolution or reconfiguration; in high dynamicity of the nowadays markets a cluster cannot be static but has to adapt to the market conditions.

The focus of this chapter concerns the operatives management of the cluster. Therefore, the composition of the enterprises proposal in order to submit a single proposal to the customer is the main problem.

So far, the negotiation approach has been considered as a process in which suppliers can act by themselves. In this case, the AA has the possibility to choose the best supplier for each request component. On the contrary, in case a group of vendors $G \subset \{1,2,\ldots,J\}$ create a coalition the offer is given by:

- $DD^G = \max_j \{DD_i^j\}$,
- $V^G = \min_j \{V_i^j\}$,
- $P^G = \sum_j P_i^j$, for each $j \in G$ and for each $i \in \{1,2,\ldots,I\}$.

In this case, the cluster offers to the AA an assured number of final products equal to V^G. The bargaining process is the same as the above explained: the only difference is that while in the first case, the competition is against the single suppliers, in this case is against the supplier cluster.

The simplicity of the proposal computation can evidence the difference in terms of performance between cluster and no-cluster approaches.

8 Simulation Environment

The agent-based architecture proposed in the research has been implemented by developing a test environment consisting of a discrete event simulation environment able to execute all the functionalities discussed in the previous paragraphs. The environment has been developed by using open software technology entirely realized by Java Development Kit. It consists of a simulation environment, developed by using Java development kit package, able to test the functionality of the proposed approaches and to understand the related advantages and/or limits. The modeling formalism here adopted is a collection of independent objects interacting via messages. This formalism is quite suitable for Multi Agent Systems development. In particular, each object represents an agent, and the system evolves through a message-sending engine managed by a discrete event scheduler. Specifically, the agents showed in Fig. 11 have been developed.

In particular, those agents have the following tasks:

- *schedulator agent*; it activates the other agents and manages/coordinates the actions of the simulation;
- *model supplier*; it supervises the supplier local data and its actions;
- *negotiation agent (both customer and supplier);* it manages the negotiation protocol by its algorithms and provides this information to model in order to negotiate with the customer or supplier.
- *selection agent;* If a queue of orders arrives to the supplier, it selects the first order to negotiation by its objectives to reach.
- *Production Planning agent;* It performs the ERP of the supplier in order to plan the requested order; therefore, it performs an opportune interface package with

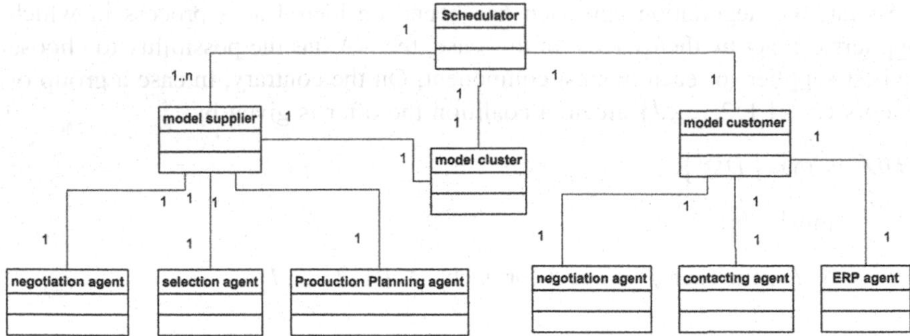

Fig. 11 Class diagram

the ERP software. Moreover, it provides the planning information to the model supplier.

- *model customer*; it supervises the customer local data and its actions;
- *contacting agent*; it supervises the supplier's database and provides this information to select the suppliers to negotiate.
- *ERP agent*; it supervises the ERP software of the customer. From the customer point of view, it updates the information of the ERP after a negotiation end with agreement. It, also, performs an opportune interface package with the ERP software.

The Architecture developed leads to the following advantages:

- Agent technology is an enabling technology to approach and solve complex distributed problems characterizing network enterprises;
- Discrete event simulation is a powerful tool for designing and testing distributed environment based on agent technology;
- Using open source technologies at the simulation package implementation phase allows code reusability for real platform development; in fact, the actual platform package can be built by utilizing the same code and architecture used for the simulation environment allowing time and investment cost saves. This reduces the risk related with investment in ICT and agent-based technology in distributed production planning.

9 Test Case

The assumption is that the AA has the possibility to make and sell three product typologies:

- a "classic bike", code I100;
- a "mountain bike", code I200;
- a "velocity bike", code I300.

Table 1 Assembler's suppliers

Components	Suppliers
Frame	1, 2, 3
Wheel	4, 5, 6
Speed gear	7, 8, 9
Accessories	10, 11, 12
Brake	13, 14, 15

The considered suppliers are 15 (J), three for each component: Frame, Wheels, Speed Gear, Accessories, Brake, like specified in Table 1. The Buyer Agent requests 15 orders, five for each product typologies. Table 2 shows the 15 orders received by the assembler, in terms of volumes and remaining days to reach the due date respect to the current date (denoted by "Days"). Also, each supplier has five typologies of work-stations to produce the required goods; these work-stations are indicated in the model as WS1, WS2, WS3, WS4 and WS5. Table 3 shows the routing of the components, equal for each seller. The table values indicate the standard throughput, expressed in *items/hour*, of a resource with respect to a single typology of component.

In the bargaining cluster case the seller groups G that have been created to compete are showed in Table 4. Furthermore, the following values have been assumed in the test case: $c = 1$, $c_o = 1{,}1$ and $n = 7$.

10 Results

Simulation experiment results are reported in Table 5, where the average utility (AU) for the Assembler Agent is calculated as it follows:

$$AU = U_v + U_p + U_{dd}, \tag{7}$$

being:

U_v = *Requested Volume/Obtained Volume;*

U_p = *Requested Price/Obtained Price;*

U_{dd} = *(Requested Due Date—Order Insertion Date)/(Obtained Due Date—Order Insertion Date);*

Also, the *Total Profit* is the seller's profits sum, while the *Profit Uniformity* is the difference between the maximum and minimum seller profit values, finally the total supplied volume by the customer is reported.

As the reader can notice, the AU is minor in case of no cluster; in case of cluster the negotiation concerns all the cluster; therefore, the assembler selects the best cluster, while in case of no cluster the assembler selects the best supplier for each item.

The total profit increases because in case of cluster; this is caused by the major contractual power of the cluster compared to a single supplier.

Table 2 Orders received by the assembler

Requested product			Order 1	Order 2	Order 3	Order 4	Order 5
I100		Volume	9100	7880	4890	4980	3220
		Days	5	6	7	8	9
BOM explosion	I101	Volume	9120	7890	4890	4980	3240
		Days	3	4	5	6	7
	I102	Volume	18240	15780	9780	9960	6480
		Days	3	4	5	6	7
	I103	Volume	9120	7890	4890	4980	3240
		Days	3	4	5	6	7
	I104	Volume	9120	7890	4890	4980	3240
		Days	3	4	5	6	7
	I105	Volume	9120	7890	4890	4980	3240
		Days	3	4	5	6	7
I200		Volume	8950	6720	5490	4410	2780
		Days	5	6	7	8	9
BOM explosion	I201	Volume	8950	6725	5500	4425	2800
		Days	3	4	5	6	7
	I202	Volume	17900	13450	11000	8850	5600
		Days	3	4	5	6	7
	I203	Volume	8950	6725	5500	4425	2800
		Days	3	4	5	6	7
	I204	Volume	8950	6725	5500	4425	2800
		Days	3	4	5	6	7
	I205	Volume	8950	6725	5500	4425	2800
		Days	3	4	5	6	7
I300		Volume	9120	7890	4890	4980	3240
		Days	5	6	7	8	9
BOM explosion	I301	Volume	8300	6850	5175	5475	2800
		Days	4	4	5	6	7
	I302	Volume	16600	13700	10350	10950	5600
		Days	4	4	5	6	7
	I303	Volume	8300	6850	5175	5475	2800
		Days	4	4	5	6	7
	I304	Volume	8300	6850	5175	5475	2800
		Days	4	4	5	6	7
	I305	Volume	8300	6850	5175	5475	2800
		Days	4	4	5	6	7

The total supplied volume reduces in case of cluster because the assembler selects the best cluster, but in a cluster can be present a supplier with minor throughput capacity.

Finally, the profit uniformity is clearly improved in case of cluster. The difference between the maximum and minimum seller profit values is reduced of 50 %. This is the main motivation to formation a cluster for the supplier.

Table 3 Components routing

Components	Routing				
	ws1	ws2	ws3	ws4	ws5
Frame	150	0	110	0	130
Wheel	100	103	0	180	0
Speed gear	0	110	130	105	0
Accessories	200	0	0	0	150
Brake	0	120	190	135	155

Table 4 Clusters composition

Clusters	Suppliers
C1	1, 4, 7, 10, 13
C2	2, 5, 8, 11, 14
C3	3, 6, 9, 12, 15

Table 5 Tests results

	AU	Total profit	Total supplied volume	Profit uniformity
No cluster	2.83	75.6	555140	15
With cluster	2.73	76.69	534565	7.5

Tables 6 and 7 show the suppliers profit and volume in cluster case (Table 7) and the situation in which is not possible to constitute seller coalitions (Table 6). In case of no-cluster, for three components (frame, accessories and brake) only one supplier signs all the contracts with the supplier; for the two components (wheel and speed gear), two of three suppliers sign contracts.

In case of cluster, all the suppliers sign at least one contract, therefore, the cluster leads to distribute the profit among suppliers.

Reassuming, in both analyzed cases all the orders are satisfied. The AA utility is better when the suppliers compete lonely to reach an agreement because the negotiation protocol lead to the stipulation contract only with the best one, for each component. Also, while concerning the profit and exchanged volumes variation the two models do not present significant differences, the very differentiation between

Table 6 Tests results no-cluster

	Supplier 1	Supplier 2	Supplier 3	Supplier 4	Supplier 5
Profit	0	0	15	0	6
Volume	0	0	84890	0	47440
	Supplier 6	Supplier 7	Supplier 8	Supplier 9	Supplier 10
Profit	9.6	0	11	4	0
Volume	164160	0	68610	17240	0
	Supplier 11	Supplier 12	Supplier 13	Supplier 14	Supplier 15
Profit	0	15	0	0	15
Volume	0	87120	0	0	85680

Table 7 Tests results cluster

	Supplier 1	Supplier 2	Supplier 3	Supplier 4	Supplier 5
Profit	8	1	6	8.5	1.05
Volume	40235	5475	39180	82200	19200
	Supplier 6	Supplier 7	Supplier 8	Supplier 9	Supplier 10
Profit	6.7	8	1	6.44	8
Volume	93600	39835	5475	36565	40385
	Supplier 11	Supplier 12	Supplier 13	Supplier 14	Supplier 15
Profit	1	6	8	1	6
Volume	5475	41260	40385	5475	39820

the two approaches rely on the obtained profits distribution among the suppliers. Indeed, in the clusters case all the sellers gain some percentage of supply: this is not true when coalitions are not allowed. In this last case, only 7 suppliers acquire an order. The research has proposed an approach based on Multi Agent System and Negotiation to integrate ERP system in a Supply Chain perspective; the proposed approach does not require any variation on actors ERP, since the agent architecture constitutes an interface also among different software legacies. The proposed approach is able to support and integrate supply chain also in a partner perspective. However, partnership results have been obtained without considering any coordination policy during the cluster formation. This aspect and an appropriate methodology for clusters constitution will be object of future research.

11 Summary and Conclusion

The research presented concerns the design, development and tests an agent-based architecture able to support transaction, information sharing and exchange and even collaboration in a manufacturing enterprise network. The agent-architecture is able to support the negotiation among the customer and its suppliers in an electronic Business to Business context. It provided an ERP based on ACCESS database to customer and suppliers in order to integrate the production-planning problem in the research. The Multi Agent Architecture has been designed by the UML activity diagram. Then, two cases are described: the first one each supplier competes with others to sign the agreement with the customer; in the second case, the suppliers constitute a cluster. A cluster is a coalition among the suppliers that supply all the items necessary to assembly the final product. In order to test the proposed approach, a simulation environment has been developed of the architecture. A case study has been simulated in order to evidence the difference between the two configurations of the network.

The main contributions of the chapter are:

- A multi agent architecture has been proposed to support an enterprise network. The agent-based approach leads to several advantages. Classical advantages of the Multi Agent systems such as: efficiency, reliability, extensibility, robustness, maintainability, responsiveness, flexibility, reuse, adaptability, scalability and so on. The original advantage of the proposed agent-based architecture is the integration of enterprises in a network in which each enterprise can use a different ERP software package. In an enterprise network, often, each enterprise utilized its ERP software and there are more problems to integrate and communicate each other. This situation leads to discourage the participation of an enterprise in a network. The agent-based architecture able to interface the enterprise independently by the ERP software, each enterprise has to be performed the same protocol information through the agent of the architecture.

- A cluster enterprise network is proposed; the simulation environment evidences the advantages of this approach benchmarked to a network with no cluster. The main advantage of the suppliers is the uniformity of profit distribution among them. The profit uniformity can be a disadvantage for a supplier more competitive, this condition must be avoided in the phase of selection partners of the cluster. Moreover, from the point of view performance the simulation results evidence that the supplier gain major advantage for the customer in cluster case. A deeply investigation and opportune investigation on negotiation protocol can balance the advantages among suppliers and customer.
- The possibility to build an enterprise network with open-source resources as JAVA used to implement the multi agent architecture and, therefore, at low costs of the development of real applications of IT systems. In this chapter, ACCESS is used to implement ERP, but this can depend on single enterprise.
- A negotiation protocol is proposed to coordinate the multi agent architecture and how the proposal can be composed in case of cluster.
- The research suggests how, through simulation, to evaluate the real value of planning and negotiation tools in electronic business environment and who, among customers and suppliers, get the main advantages from them, and, therefore, should pay for them.

The proposed approaches can be extended to other applications concerning the electronic marketplaces, the concept of virtual organization and distributed production systems. The main barrier is the difficult to evaluate the real benefits obtained by these tools; the simulation tool can help to overcome this obstacle.

12 Future Developments

The directions for future research in this area are the following:

- The life cycle of the cluster. In this chapter, the configuration of the cluster is not analyzed, therefore a negotiation among suppliers to create the cluster will be analyzed. The selection phase is more important to obtain a homogeneous cluster and therefore with no enterprise penalized. Moreover, the dynamicity of the market leads to change the partners of the cluster; therefore, this negotiation approach is, also, important during the operational activity of the cluster. The methodologies to be evaluated can be Pareto analysis and multi attribute ranking in order to qualify the potential partners.

- A strategies and tactics of customer and suppliers will be proposed in order to obtain an enterprise that chooses specific strategies depending on specific objectives. A knowledge base has to be developed for each opponent in negotiation approach and it can select dynamically the best strategies in order to pursue its objectives. The methodologies to investigate are negotiation strategies with asymmetry information, fuzzy rules to select the strategies and genetic algorithms to optimize the strategies.
- The extensions of the environment to a many customers in order to obtain a many to many negotiation environment. In this case, also, the customer has to compete with other customers to sign a contract with the suppliers.
- A real case study application will be proposed in order to evidences the real value added by the proposed architecture and negotiation protocols.

References

1. Kasarda, J., Rondinelli, D.: Innovative infrastructure for agile manufacturers. Sloan Man. Rev. 73–82 (1999) (Winter)
2. Kalakota, R., Whinston, A.B.: Frontiers of Electronic Commerce. Addision-Wesley, Reading, (1996)
3. Pepper, D., Rogers, M.P.: Enterprise One to One: Tools for Competing in the Interactive Age. Double Day, New York (1999)
4. Gleickabo, J.: Faster: The Acceleration of Just About Everything. Pantheon Books, New York (1999)
5. McAfee, A.P.: The Impact of Information Technology on Operational Effectiveness: an Empirical Investigation. Harvard Business School, Working Paper, Cambridge, MA (1998)
6. Kelle, P., Akbulut, A.: The role of ERP tools in supply chain information sharing, cooperation and cost optimization. Int. J. Prod. Econ. **93–94**, 41–52 (2005)
7. Turowski, K.: Agent-based e-commerce in case of mass customization. Int. J. Prod. Econ. **75**, 69 –81 (2002)
8. Perrone, G., Renna, P., Cantamessa, M., Gualano, M., Bruccoleri, M., Lo Nigro, G.: An agent based architecture for production planning and negotiation in catalogue based e-marketplace. In: 36th CIRP—International Seminar on Manufacturing Systems, pp. 47–54, Saarbruecken, Germany, 03–05 June 2003
9. Cantamessa, M., Fichera, S., Grieco, A., La Commare, U., Perrone, G., Tolio, T.: Process and production planning in manufacturing enterprise networks. In: proceedings of the 1st CIRP Seminar on Digital Enterprise Technology, Durham, UK, pp. 187–190 (2002)
10. Ash, C.G., Burn, J.M.: A strategic framework for the management of ERP enabled e-business change. Eur. J. Oper. Res. **146**, 374–387 (2003)
11. Rudberg, M., Olhager, J.: Manufacturing networks and supply chains: an operations strategy perspective. Omega **31**, 29–39 (2003)
12. Cavalieri, S., Cesarotti, V., Introna, V.: Multiagent model for coordinated distribution chain planning. J. Org. Comp. Elect. Com. **13**(3 and 4), 267–287 (2003)

13. Kang, N., Han, S.: Agent-based e-marketplace system for more fair and efficient transaction. Decision Support Syst. **34**,157–165 (2002)
14. Argoneto, P., Bruccoleri, M., Lo Nigro, G., Noto la Diega, S., Perrone, G., Renna, P.: Evaluating multi-lateral negotiation policies in manufacturing e-marketplace. In: Proceedings of the 37th CIRP—International Seminar on Manufacturing Systems, Budapest, Hungary (2004)
15. Argoneto, P., Bruccoleri, M., Lo Nigro, G., Perrone, G., Renna, P., Sabato, L.: Integrating ERP systems in vertical supply chain with negotiation tools. In: EurOMA International Conference on Operations and Global Competitiveness, Budapest, Hungary, 19–22 June 2005. ISBN 963-218-455-6
16. Perrone, G., Bruccoleri, M., Renna, P.(eds.): Design and Evaluating Value Added Services in Manufacturing e-marketplace. Springer, Berlin (2005).ISBN: 1-4020-3151-3
17. Lea, B., Gupta, M.C., Wen-Bin, Yu, W.: A prototype multi-agent ERP system: an integrated architecture and a conceptual framework. Technovation, **25**(4), 433–441 (2005)
18. Xu, L., Beamon, B.M.: Supply chain coordination and cooperation mechanisms: an attribute-based approach. J. Supply Chain Man. **42**(1), 4–12 (2006) (winter)
19. Gunasekaran, A., Lai, K., Cheng, T.C.E.: Responsive supply chain: a competitive strategy in a networked economy. Omega, **36**(4), 549–564 (2008)
20. Renna, P.: Negotiation policies and coalition tools in e-marketplace environment. Comp. Ind. Eng. **59**, 619–629 (2010)
21. Renna, P., Argoneto, P.: Production planning and automated negotiation for SMEs: an agent based e-Procurement application. Int. J. Prod. Econ. **127**, 73–84 (2010)
22. Saeed, K.A., Malhotra, M.K., Grover, V.: Interorganizational system characteristics and supply chain integration: an empirical assessment. decision sciences **42**(1), 7–42 (2011)
23. Olson, D.L., Staley, J.: Case study of open-source enterprise resource planning implementation in a small business. Ent. Inf. Syst., **6**(1), 79–94 (2012)
24. Shafiei, F., Sundaram, D., Piramuthu, S.: Multi-enterprise collaborative decision support system. Expert Syst. Appl. **39**(9), 7637–7651 (2012)
25. Nazemi, E., Tarokh, M.J., Djavanshir, G.R.: ERP: a literature survey. Int. J. Adv. Man. Tech. **61**, 999–1018 (2012)
26. Chen, K.: Procurement strategies and coordination mechanism of the supply chain with one manufacturer and multiple suppliers. Int. J. Prod. Econ. **138**(1), 125–135 (2012)
27. Presley, A., Sarkis, J., Barnett, W., Liles, D.: Engineering the virtual enterprise: an architecture-driven modeling approach. Int. J. Flex. Man. Syst. **13**, 145–16 (2001)
28. Feldman, G.C.: The Practical Guide to Business Process Reengineering Using IDEF0. Dorset House Publishing, New York (1998)
29. Porter, M.: The Competitive Advantage of Nations. New York, The Free Press, New York (1980)
30. Schmitz, H.: On the clustering of small firms. IDS Bull. **23**(3), 64–68 (1992)
31. Porter, M.: Clusters and the new economics of competition. Harv. Bus. Rev. **76**, 77–90 (1998)
32. Steiner, M., Hartmann, C.: Learning with clusters: a case study from Upper Styria. In: Steiner M (ed.) Clusters and Regional Specialisation—On Geography, Technology and Networks, vol. 8, pp. 211–225. European Research in Regional Science (1998)

Chapter 7
Adaptive Applications: Definition and Usability in IT-based Service Systems Management

Ammar Memari and Jorge Marx Gómez

Abstract Adaptive applications are not well defined in literature, and are usually confused with other sorts of applications. In this chapter we try to capture common properties of these applications into an informal definition. Then later come up with a formal definition on the basis of a comprehensive set of attributes. This domain-independent definition of adaptive applications is then employed and exemplified in the field of IT-based Service Systems Management.

Keywords Adaptive applications · Rough sets · Definition · Adaptive hypermedia · IT-based service systems

1 Introduction

Adaptive applications as a subclass of intelligent systems have attracted lights recently due to their abilities to optimize resource consumption, increase user satisfaction, decrease maintenance costs through automatic adaptation to new circumstances, and aid coping with massive amounts of information through automatic filtering.

The term "adaptive application" is however used nowadays in ambiguous ways referring to a wide spectrum of applications. Efforts to strictly define this class of applications are not fruitful due to the fuzzy nature of it, which gives different applications different membership degrees to this class.

A. Memari (✉) · J. Marx Gómez
Business Informatics I/Very Large Business Applications, Carl von Ossietzky University of Oldenburg, Ammerländer Heerstr. 114-118, 26129 Oldenburg, Germany
e-mail: ammar.memari@uni-oldenburg.de

J. Marx Gómez
e-mail: jorge.marx.gomez@uni-oldenburg.de

M. Mora et al. (eds.), *Engineering and Management of IT-based Service Systems*, Intelligent Systems Reference Library 55, DOI: 10.1007/978-3-642-39928-2_7,

In this book chapter we try to define this class along with other neighboring classes using the concept of Rough Sets [1] as an instrument for specifying such a blurry class of applications.

The definition will later serve provide a clear view on how adaptivity would support management of service systems by automatic discovery and recommendation of more relevant and better-performing services.

2 A Rough-Sets-based Definition

In order to define the set of adaptive applications, we will approach it as a rough set where each application has a membership degree corresponding to the features it holds.

2.1 Set of Applications Under Study

We choose a set of applications *O* familiar to the reader so the definition of the various attributes could be clearer. Concrete applications under study are: Amazon recommendation engine,[1] Last.fm,[2] Facebook ads engine,[3] Waze social navigation application,[4] and Jinengo multimodal transportation planner.[5] In addition to two classes of applications: A typical Adaptive Hypermedia (AH) application and a typical load balancer.

2.2 Rough Sets

According to [1]: A rough set is a formal approximation of a set in terms of a pair of sets which give the lower and the upper approximation of the original set. Where the upper approximation is the complete set of objects that are *possibly* members of the target set and the lower approximation is the complete set of objects that *definitely* belong to target set.

In this chapter we will study the set *A* of attributes that an application might have in relevance to adaptivity, and a set of applications *O* that are familiar to the reader will be under study and examination from the attributes point of view.

---

1 http://www.amazon.de

2 http://www.last.fm

3 http://www.facebook.com

4 http://www.waze.com

5 http://www.jinengo.com

2.3 Definition of Lenz

Lenz [2] differentiates three levels of details for addressing and distinguishing users:

Typification: Addressing information to a general user group, without regard to the single receiver.

Individualization: Addressing single users or typical user groups, still not regarding the concrete person.

Personalization: Identification of single persons that the system needs to deal with.

Moreover, Lenz categorizes the extent to which the adaptation of information to the user is possible into three categories:

Adapted systems: Represent the easiest constructs from the flexibility point of view. Adaptation to the user takes place only during implementation time, and it is usually oriented towards a fixed, homogeneous target group

Adaptable systems: Allow the user to change the system and to adapt it according to his needs. That would guarantee the system larger flexibility (on the expense of higher complexity) allowing it to be used by a heterogeneous target group.

Adaptive systems: Take over adaptation to a specific user group/user profile autonomously and adapt themselves to current needs without assistance. They have the advantage of simplified usage, but can possibly lead to lack of transparency and controllability of the adaptation from the user side [2]. Furthermore, adaptive systems require sophisticated control to automatically determine the appropriate requirements of the users. Through juxtaposition of these two attributes we can clearly classify different communication systems (see Table 7.1).

We encode the definition of Lenz into an attribute set as follows:

$$P_{Lenz} = \{P_{typified}, P_{individualized}, P_{personalized}, P_{adapted}, P_{adaptable}, P_{adaptive}\}$$

This encoding does not correspond exactly to the classification of Lenz since it includes cases where an application is both adaptive and adaptable that Lenz's definition excludes. Such cases should not be ignored and we will see examples for them in the upcoming sections of this paper. Some applications have a layer of adaptivity on the first order, and adaptability/adaptedness on the second order. These are also out of scope for Lenz's definition, and we will discuss them further ahead.

Table 7.1 Communication systems classification according to [2]

Level of user modeling	Personalized			
	Individualized			
	Typified			
		Adapted	Adaptable	Adaptive
		Level of adaptation		

2.4 Definitions in Adaptive Hypermedia

Researchers in the field Adaptive Hypermedia have distinguished also between adaptive and adaptation systems through the definition from Goy et al. [3] which went along with De Bra et al. [4] and Brusilovsky [5] and is based on Oppermann [6] as follows:

> "In adaptable systems the adaptation is decided by the user, who explicitly customizes the system to receive a personalized service."

> "In adaptive systems, the adaptation is autonomously performed by the system, without direct user intervention."

However, research in the Adaptive Hypermedia focused mainly on the storage layer rather than the application layer. This is justifiable as the field discusses Hypermedia not applications on top of Hypermedia.[6] The focus of the Dexter Hypertext Reference Model is on the storage layer, as well as on the mechanisms of anchoring and presentation specification that form the interfaces between the storage layer and the within-component and runtime layers [7]. So inherently, and since major AH reference models are Dexter-based, they have also taken little notice of the runtime/application layer.

3 Set of Attributes Relevant to Adaptivity

Here we list the attributes of set *A* under which we will be studying our set of applications *O*.

3.1 Models Adaptation Subject

Adaptation subject is the answer to the question: "To what to adapt?" according to the LAOS model suggested in [8]. However, according to that model, only the application user is considered to be the subject. Here we expand this concept to include besides the user three other adaptation subjects:

3.1.1 Models User as Adaptation Subject

The user is the key subject in the adaptation process, especially in the research field Adaptive Hypermedia, where the term user conventionally represents the

6 In Adaptive Hypermedia terminology the applications layer is termed as the runtime layer.

human end user dealing directly with the application under study. However, the user in our case can be a software agent, or a calling service in a Service-Oriented Architecture. Sticking to the definition of Lenz [2], depth of user modeling is on three levels:

Typified: Addressing information to a general user group, without regard to the single receiver [2]. For example: setting up a webpage for English speakers.

Individualized: Addressing single users or typical user groups, still not regarding the concrete person [2]. For example a Website detects the IP of the visitor and changes language based on IP location. In this case even though the application addressed a single user, it did not adapt to more than one attribute of him (location) which can define a group of users.

Personalized: Addressing single persons whom the system needs to deal with [2]. For example, applications requiring login like Amazon.de or Last.fm where a concrete person is addressed by a set of attributes that define him particularly such as a unique email address or user name. In addition to the specific person's purchase history or music listening history.

3.1.2 Models Neighborhood as Adaptation Subject

Modeling and adapting to other users of the application is what we call the neighborhood adaptation. Mostly well-known for Collaborative Filtering where if user *x* shares the same attributes *a1*, *a2*, *a3* with user *y* then there is a high probability that they also share the attribute *a4*. For example, Amazon would recommend to you to buy an item that other users who share a considerable deal of purchase history with you have bought. Here the Neighborhood model is based on the user-content "buys" relation. However, in other cases it could be built on other relations like when your Facebook friend likes an article it gets recommended to you even though you have no common reading history, but merely because you are Facebook friends, a user–user "is a friend of" relation. More about different relations can be found in Sect. 3.3.

3.1.3 Models Content as Adaptation Subject

Content is usually thought of as an adaptation object rather than a subject. Nevertheless, content is an important subject of adaptation. In Last.fm for example, where the different songs are the content, modeling content relationships such as "same genre" and "same artist" is crucial for the adaptation process.

Content as a subject and an object for adaptation is usually represented by a limited set of items and relations among them; the typical example in the field of Adaptive Hypermedia is an educational curriculum. All items of the curriculum are known to the application beforehand; they are annotated by the author according to their interdependency, difficulty, and other attributes in authoring time completely before running the application. Such a corpus of information is referred to as a "closed corpus" [9]. On the contrary, when content items are not known beforehand and are susceptible to change, we then have an "open corpus". Open corpus content is a challenge for conventional adaptive applications that require a pre-structured pre-indexed set of items.

Crawling the open corpus is one of the many issues this nature of the content imposes (see [10]). Jinengo application is a good example for dealing with the open corpus where new train connections, altered bus routes and new flight destinations are crawled and added autonomously to the content repository by intelligent crawlers working in a beehive-like manner (see [11]).

3.1.4 Models Context as Adaptation Subject

In many occasions, applications need to adapt to some factors that are not directly related to the user's preferences; these factors is what we call context. For example, a trip planner like Jinengo needs to check the weather at the time of the trip, and when rain is expected, it has to keep the bike away from the list of transportation possibilities. Other examples for context include device screen size, connection speed, and connection type which are of high importance for Mobile Computing [12, 13].

A load balancer distributing requests among a set of servers is able to sense how utilized each server is. Even though it does not model the user whatsoever, it still has an adaptation subject which is the context.

3.2 Models Adaptation Object

Adaptation object is the answer to the question: "what to adapt?" [8]. The author in [14] did not distinguish between content adaptation and presentation adaptation; the author considered both to be presentation adaptation which is separate from navigation adaptation (Link-level adaptation). However, in later works of Adaptive Hypermedia we can see that the separation is clearer: while content adaptation means delivering different information to different users (e.g. based on different knowledge levels), presentation adaptation means presenting the *same* information differently (e.g. same information, different language).

3.2.1 Models Content as Adaptation Object

Content is both a subject to adaptation as discussed in Sect. 3.1.3 and an adaptation object. A conventional educational AH application for instance filters out information that have a knowledge level higher or lower than the user's.

3.2.2 Models Presentation as Adaptation Object

Different presentation of the same information based on an adaptation subject is what's meant by presentation adaptation. For example, Waze uses different display colors at night time based on sensed time of day (context); thereby presenting the same navigation information differently. Most of mobile-enabled websites deliver a different version of the site when they sense a mobile device (context). Creation of this version involves content adaptation by removing elements that heavily consume resources.

3.2.3 Models Navigation as Adaptation Object

Navigation means traversing the different "pages" of an application by different means. On the Web navigation is accomplished mainly through web links, and in enterprise applications buttons and menus allow the user to traverse from one screen (transaction) to another.

Adaptive navigation helps users find their paths in the hyperspace [14]. For example, a web browser can use the browsing history of the user to change the color of links leading to already visited pages thereby giving a hint to the user who changes his navigation path accordingly. Last.fm adapts navigation by selecting the next song to be played after the user clicks on "next" or when the current song ends.

3.2.4 Models User as Adaptation Object

Even though the user is usually the main adaptation subject, it is sometimes an object as well. Applications that alter the user model based on her actions learn the user model implicitly by interacting with her. Traditionally, user modeling (building the user model) was completely separated from adaptation (adapting to user model). This can be clearly seen in the first figure in [14]. Nevertheless, recent applications tend to merge the two together by having the user model as an adaptation object.

For example, Last.fm learns about the user preferences from songs he listens to, genres or artists he searches for, and songs he enlists on his "loved" list. Last.fm updates the user model implicitly based on these events without an explicit intervention from the user. The same can be said about Waze, it implicitly keeps a list of "favorite" paths between specific destinations simply when the user drives these paths frequently. It then uses this list in the routing algorithm.

3.3 Models Where and How to Adapt

Adaptation model represents where and how to adapt, it is usually a compilation of adaptation rules. These rules define the adaptation, by combining the subject and object; sometimes it is also called the teaching model [15, 16]. If we map this model to if–then rules, the "if" part of the rule determines *where/when* adaptation should take place and the "then" part determines *how*.

3.3.1 Models Adaptation Condition (Where)

Adaptation conditions depend on relations between adaptation subjects. Theoretically all combinations of the four subjects are possible resulting in $C_3 = 6$ binary relations, $C_3 = 4$ ternary relations and 1 quaternary relation. So theoretically there are 11 types of relations that could be modeled. However, not all combinations are utilized by applications under study therefore we will focus on the most frequently used ones.

Matches User to Content

The application utilizes a relationship between the user model and the content model. Examples for such relations are many: the relationship "bought" in Amazon, "listened to" in Last.fm, "travelled (a path)" in Jinengo or Waze and "studied (a paragraph)" in an educational AH application. In many cases this type of relation alone is not enough for an adaptation decision.

Matches Content to Content

An application finds and utilizes a relationship between two or more content items. We can find many examples for it: "belongs to the same album as" relationship between two songs on Last.fm, "belongs to the same category/price range" between two items on Amazon and "has the same origin and destination" between different paths on Waze. This type of relationship is used usually together with the user-content type to reach an adaptation decision as follows: User x "bought" item a and item a "belongs to the same category/price range as" item $b \Rightarrow$ recommend item b to user x.

Matches User to User

Application finds and utilizes matches between different users. This type of relation is dominant in "Collaborative Filtering" applications and through this type of relations a neighborhood model is formed. For example the "is friend of" relation in Facebook is utilized by the ads engine to display to a user the advertisement of a game played by one of her friends. Sometimes this type of relation is implemented using other types, for example on Last.fm and besides the direct "is friend of" relation, a user can find others who share the same musical taste, i.e. other users who have "listened to" songs "of the same genre as" songs this user had "listened to".

Matches Content to Context

Application finds and utilizes matches between content and context. A clear example is adapting websites to screen size or connection speed. The relation between content and context in this case is for example "requires connection bandwidth" or "requires screen size". When an element of a page requires a high connection bandwidth it would be filtered out if the current connection is not of a high bandwidth.

Matches User to Context to Content

Application utilizes a ternary relation between these three aspects. Ternary relations are found not so often in applications, an example of this relation is found in Jinengo where the application would consider the rainy weather (context), the bike path (content) and the user preference "likes riding the bike in a rainy weather" (user) to reach an adaptation decision whether to include this path in the search results.

3.3.2 Models Adaptation Action (How)

The adaptation effect can vary largely depending on the application. Adaptation can affect, as mentioned before, content, presentation and navigation. Here we list the most significant manifestations of adaptation effects [17].

Inserts/Removes Fragments

Parts of the page are added or removed. For example, adding a personalized advertisement to a web page.

Alters Fragments

For example an AH application changes the explanation of an idea to use a simpler language for novice users.

Dims Fragments

For example when one of the resulting routes in Jinengo is very harmful to the environment; it is dimmed in the result list.

Sorts Fragments

The result list of different routes in Jinengo is sorted adaptively based on the user model.

Uses Stretchtext

Stretchtext is a technique that enables the application to expand/collapse a certain item without having to move to a new page. In Jinengo, when one route on the result list is clicked, it expands showing details without changing the page, however, that is not done automatically but rather through user explicit interaction.

Zooms/Scales

In order to give higher focus on what is relevant to the adaptation subject, a zooming/scaling operation could be done by the application:

- Uses Conventional Scaling/Zooming: Waze zooms the map based on the current speed or the distance left to reach the destination. Mobile device web browsers zoom on the clicked area when the click hits two or more links at the same time in order to allow the user to be more specific.
- Uses Fisheye View: Fisheye view gives detailed description of relevant items while keeping irrelevant ones less detailed or even as an outline. Thereby keeping the focus on the most relevant without losing the big picture on the whole.
- Summarizes Fragments: Fragments that have less relevance are displayed as a summary rather than being collapsed or removed.

Adapts Layout

Certain layouts are more appropriate under certain circumstances. Mobile devices have smaller screens therefore a different page layout is more appropriate, many web applications have a mobile version featuring a mobile-friendly layout.

- Uses Partitioning/Zooming: Mobile version of Facebook cuts off profile pictures and displays other photos in a lower resolution.
- Uses Rearrangement: Elements on the mobile version of Google search engine are arranged differently to better suit a smaller screen and better serve the purpose of a quick search.
- Fits Layout to a Template: Some websites feature "themes" allowing the user to change colors, borders, fonts and other display characteristics of pages. Thereby fitting the same information into a different template.

Sorts/Orders Links

To control navigation, applications change the order of links. For example, an enterprise portal would sort the links in a way that keeps mostly-used tasks of the user on top.

Annotates Links

Simplest form of link annotation is changing the color of links leading to pages visited before; this is a default behavior in web browsers. Empirical studies of adaptive annotation in the educational context have demonstrated that it can help students acquire knowledge faster, improve learning outcomes, reduce navigational overhead, and encourage non-sequential navigation. Moreover, [18] shows that it can significantly increase students' motivation to work with non-mandatory educational content.

Uses Combinatorial Techniques

We distinguish here four kinds of links that are usually used for navigation in applications:

- Uses Contextual Links: Links embedded within elements of the page like a hot spot in a picture or a hot word in a text. They can be annotated but cannot be sorted or hidden [14].
- Uses Non-Contextual Links: Links independent from content of page.
- Uses Contents/Index Links: Links on an index and contents page have a fixed order, but are still annotatable.
- Uses Local and Global Maps: Maps are a way to graphically represent a hyperspace. For example site map.

Generates Links

When viewing an item at Amazon, links to related items are generated. If the user is recognized, eBay would generate links on the home page leading to items relevant to the search and purchase history of the user.

- Adapts Anchor: Anchor is the "hot" point on the page, enabling and disabling anchors is a way to provide adaptive navigation leading to explanations of some terms for example.
- Adapts URL: Adapting the URL of a link changes the link's destination allowing it to lead the user to a page or resource more relevant to her or to the adaptation subject.
- Adapts Destination: The difference from URL adaptation is that with the former the decision as to which link destination to use is made when the link is generated, whereas the latter always shows the same URL, but when the link is accessed the server decides which actual destination to return. For example the "Next" button in Last.fm does not lead to a fixed destination but the destination song is decided by the server based on an adaptation decision.

Guides

The application guides the user to make navigation decisions:

- Guides Locally: Directs the user on how to take one navigational step starting from the current page.
- Guides Globally: Guides the user throughout multiple pages based on the user's goal.

Hides Links

Hiding irrelevant links helps the user not to get lost in the hyperspace or distracted from the goal.

- Hides Links: Completely hides the link, used mainly for non-contextual links that can be hidden without affecting contents of the page.
- Disables Links: When links are contextual, disabling them keeps the page content intact and serves the purpose of preventing distraction.
- Removes Links: Removing a non-contextual link entirely from the page does not only remove it from view but also can save processing resources used for generating the link's URL.

3.4 Has Level and Order of Adaptation

We divide according to [2] adaptation into three levels: adapted, adaptable and adaptive. We also define the higher order adaptation as being adaptation of adaptation, in other words, adaptation that has the adaptation process as the object.

3.4.1 Has First-Order Adaptation

First order adaptation is the conventional adaptation where adaptation arguments (subject, object, condition, action…) cannot be adaptation functions. The three levels of adaptation correspond to Lenz's definition in Sect. 4.

Is Adapted on the First Order

An adapted application is adapted in implementation time. For example, a social network web application targeting business people or academics like Xing.com or LinkedIn.com. Such applications are adapted for these user groups inherently from the design time.

Is Adaptable on the First Order

An adaptable application is manually changeable by the user according to her needs; thereby allowing the user flexibility in some aspects. Most of the applications are adaptable in some sense since most of them provide the ability for the user to change some settings. The user model in this simplistic view is the combination of settings the user sets even if they don't contain any personal information.

For example, if we have a simple online flash game that has two settings for toggling music and game sounds on or off, the game sees four models of users according to the four possible combinations of these settings.

Is Adaptive on the First Order

There is no clear border between adaptability and adaptivity; nevertheless we can stress some distinguishing characteristics: Autonomy and implicitness in sensing the adaptation subject. Behavior of the user on Last.fm, Amazon or an educational AH application is tracked down and saved to be acted upon later. Some contexts such as weather could be found out automatically using web services in Jinengo and without the user's request. Autonomy in updating the subject model: Traveling a certain path many times would alone alter the user's model in Waze without the user's explicit intervention. Learning a section in an educational AH application would result in an *automatic* update of the user model accordingly. The user is not notified or consulted about these changes. Reuse of the subject model. Amazon's recommendation engine, for example, uses the user model on every product page the user visits and on the home page, and Facebook ads engine uses the Neighborhood model every time before displaying advertisements.

3.4.2 Has Second-Order Adaptation

Adaptation of adaptation is accomplished by having a flexible adaptation function. This function itself can be altered to better suit a second-order adaptation subject. The adaptation function is in this case the second-order adaptation object.

Is Adapted on the Second Order

Due to the fact that the first-order adaptation function is defined on design time, we can consider all applications that are adaptable or adaptive on the first order to be adapted on the second order. By defining this adaptation function programmers adapt the application to be adaptable or adaptive.

Is Adaptable on the Second Order

As stated before, adaptability means ability to manually alter the application according to subject, and on the second order adaptability is the ability to manually alter the adaptation function. For example, Waze allows disabling the adaptation functionality of tracking favorite paths, thereby allowing the user to adapt the adaptivity function. Last.fm allows the user to determine how much percent of the track should play before it is considered to be "listened to", this as well is adaptability of the adaptation function.

Is Adaptive on the Second Order

Adaptivity on the second order means that the adaptation function is adapted autonomously by the application. Jinengo uses this approach by utilizing a genetic engine for crossbreeding best performing agents to produce a potentially better generation. Important to mention that the evolution process is guided by user's implicit feedback therefore we can say that this process is an adaptation process, and it is an adaptation of adaptation agents and adaptation function [11].

3.5 Has Navigable Conceptual Space on Models

A conceptual space allows aggregation and navigation in the space of concepts and not mere pages. In Amazon for example items are sorted into categories and subcategories. When a user interacts with an item, the system registers user interest not only in the item but in its category as well. Hierarchies on the content model are the most commonly used, however, hierarchies and more sophisticated classification techniques are applicable to all models of subjects, objects, conditions and actions.

3.5.1 Single-faceted (Hierarchy)

The simplest among classification structures; it has one root and the only allowed relation is the "is a" generalization relation between a child and its immediate parent. On subject: For example, adaptation to different levels of a hierarchy such as adaptation to European users, German users, one specific user. On object: For example, adapting content on the category level, subcategory level or single item level. On condition: Rules conditions correspond to subject hierarchies for example by matching a user to a category of items. On actions: Actions can have the hierarchy of the object, for example the hierarchical structure of a website requires the adaptation action to affect different levels of the website hierarchy.

3.5.2 Multi-Faceted

A multi-faceted classification classifies the same objects from multiple points of view (facets) based on different characteristics. Unlike the hierarchy that has one root, a multi-faceted classification has a root for each attribute. For example a track on Last.fm can be categorized according to genre, artist, length, album, release date…and each of these attributes can have an independent hierarchy. This is an example of having a multi-faceted classification on the content model that serves as a subject and an object of adaptation. In a similar manner we can find examples for a multi-faceted classification on other models like classifying the neighborhood in Facebook according to the region hierarchy or the interests hierarchy.

3.5.3 Ontology

An Ontology is not merely a multi-faceted classification since it features a much richer set of relations than the "is a" hierarchical relation. Jinengo utilizes ontologies for representing the different models of adaptation, and this allows the system a deeper understanding of the subject/user by being able to represent complex relations among the user interests. One interesting way to represent the domain using ontologies is implemented in the POWDER protocol [19]. This way is suitable for the open corpus since the description resources don't have to be centralized, so they can be crawled together with content then incorporated into the stored domain model. Ontological representation can be used for modeling adaptation conditions as we can see in the SPARQL language where complex ontological conditions can be defined for matching ontological models. Actions can also be represented in an ontological manner using POWDER's resource grouping to identify the targeted section of a website for the adaptation action.

3.5.4 Shared Ontology

Ontologies specify a shared conceptualization [20, 21]. They can be integrated with each other through inter-ontology relations in a process called ontology alignment, allowing thereby seamless data integration among different applications and assuring higher knowledge quality by allowing separation of concerns among domain experts. For example, description of Co2-emission-related knowledge in Jinego's domain model imports definitions from a global ecology ontology, and shares the transportation ontology it defines so other applications can use and refer to.

3.6 Has Adaptable Architecture

Application architecture can be adaptable, that can be implemented in applications composed of loosely-coupled modules/components. For Jinengo, loosely-coupled software agents are the composing components. Other service-oriented and components-oriented applications also fit into the adaptable architecture class since their architecture is at least manually changeable. Applications that have an adaptive architecture are ones that can alter their architecture autonomously, for example, Jinengo can autonomously replace agents based on their performance and user rating through cross-breeding and evolution [11].

3.7 Models Goal of Adaptation

The goal model, also known as "goal and constraints model", stores the aim of the application from the perspective of e.g. a teacher, as far as a learning environment is concerned. Also pedagogic information or the business logic of a commercial site could be stored in the goal model [22].

3.8 Respects Business Constraints

Business applications run within the business environment, so they affect and are affected by it. Here we list four of the restrictions the business environment imposes on adaptive applications:

3.8.1 Privacy

Most of adaptive applications store personal data about the user as a subject model. These data are market precious and legally sensitive. Adaptive applications are used within the frame of business personalization leading to profit when companies strictly care for the customer's individual privacy and don't undermine his interests, don't exploit them and don't betray his secrets [23]. An example for a technical incarnation of a privacy restriction is when personal data are legally not allowed to leave the user's machine, a mobile agent is sent to that machine instead to do the matchmaking and return with the result.

3.8.2 Scrutability

Since an adaptive application shapes the object for the user, a high level of trust must be present, so the system must be able, when requested, to explain in a human-readable language why it took the decision to recommend an item and hide another. That is represented by the concept of scrutability [24]. Enabling the user to see the application's model of his and the logic behind the application's adaptation decision are the most important aspects of scrutability.

3.8.3 Tunneling-Proofness

Another phenomenon that can appear in Adaptive applications is Tunneling, where the user's attention is directed strictly to what the application recommends whereas other potentially irrelevant information is cut off. To avoid such situations a certain level of randomness must exist in the system.

4 Adaptive Applications Definition

Let U be the universe of all applications, and $A = \{A_1, A_2, A_3, \ldots A_n\}$ be the set of application attributes mentioned in Sect. 3, and V be the set of values these attributes may take. For simplicity, we assume that $V = \{0, 1\}$ for the terminal attributes so each application either has the attribute or does not have it.

$$\forall \mathrm{a} \in \mathrm{A},\ \mathrm{a:U} \rightarrow V_a$$

For the non-terminal attributes, we will simply calculate the value to be the sum of the children values. For example, the value for attribute "Has first order adaptation" is calculated by summing the values for attributes: "Is adapted on the first order", "is adaptable on the first order", and "is adaptive on the first order".

Defining a set of applications through the set of attributes does not address single applications but rather defines equivalence classes corresponding to all possible combinations of possible attributes values: $[x]_A = \{x \in \mathrm{U} | V_a(x) = 1\}$.

Let's try to define the adaptive applications set $X \subset U$ by using the attribute subset A:

Application x is an adaptive application ($\in X$) if:

- it models at least one adaptation subject: $\forall x \in X, V_{subject}(x) \geq 1$
- it models at least one adaptation object: $\forall x \in X, V_{subject}(x) \geq 1$
- it models at least one adaptation rule (one condition and one action): $\forall x \in X, V_{condition}(x) \geq 1\ and\ V_{decision}(x) \geq 1$
- it is adaptive on the first order: $\forall x \in X, V_{1\mathrm{st}\, adaptive}(x) = 1$

X cannot be expressed exactly, because the set may include and exclude applications which are indistinguishable on the basis of attributes A. As a rough set, being exactly inexpressible is by definition, it rather can be approximated by expressing the upper and lower approximations as follows:

4.1 Lower and Upper Approximations

The *lower approximation* or *positive region* is the union of all equivalence classes ($[x]_A$) that are contained by X.

$$\underline{A}X = \{x | [x]_A \subseteq X\}$$

The lower approximation of our adaptive applications set X represents all applications that can positively be classified as belonging to the set X, thereby having a probability of 1 for belonging to the set.

The *upper approximation* or *negative region* is the union of all equivalence classes ($[x]_A$) that have a non-empty intersection with X.

$$\bar{A}X = \{x | [x]_A \cap X \neq \emptyset\}$$

In our case, this upper approximation represents all applications that have any of the attributes that define adaptive applications (see Sect. 6). In other words, the upper approximation contains objects that have a non-zero probability of being members of the set (including the lower approximation members).

Based on upper and lower approximations we can define the rough set of adaptive applications to be the tuple: $\langle \underline{A}X, \bar{A}X \rangle$.

4.2 Reduct (Core)

As we see in Sect. 6, not all attributes of set A are involved in the definition, therefore we can say that a subset of the attributes can fully characterize it. This subset is called a reduct [1]. About this set we can say that $RED \subseteq A$, and that $[x]_{RED} = [x]_A$, which denotes that the same equivalence classes are produced by RED as by the original set A.

The reduct is minimal, that means we cannot remove any attribute from it without affecting the equivalence classes: $[x]_{RED-\{a\}} \neq [x]_{RED}$. So the reduct (core) for our definition is the set containing only the effective attributes.

4.3 Other Definitions

In the same way we can define other types of applications based on the set of attributes. For example, we can slightly change the previous definition to get the definition of adaptable applications or adaptable adaptive applications by changing the last rule to become: $\forall x \in X, V_{1stadaptive}(x) = 1$, or: $\forall x \in X, V_{1\text{st } adaptive}(x) > 1$ respectively. We can also use this methodology to formalize some informal definitions. For example, we can define "Collaborative Filtering Applications" $X_{CF}$to be applications that:

- model neighborhood as adaptation subject: $\forall x \in X_{CF}, V_{neighborhood}(x) = 1$
- model content as adaptation object: $\forall x \in X_{CF}, V_{ocontent}(x) = 1$
- match user to user: $\forall x \in X_{CF}, V_{conditionU-U}(x) = 1$
- model at least one adaptation action: $\forall x \in X_{CF}, V_{action}(x) \geq 1$
- are adaptive on the first order: $\forall x \in X_{CF}, V_{1\text{st } adapptive}(x) = 1$

Similarly we can define "Context-Aware Applications" as applications that model context as adaptation subject: $\forall x \in X_{CF}, V_{scontent}(x) = 1$, and at least one adaptation condition involving context: $\forall x \in X_{CF}, V_{conditionCx-Ct}(x) + V_{conditionU-Cx-Ct}(x) + \cdots \geq 1$. The rest of the attributes don't matter in this case.

The definition of "Adaptive Applications on the Semantic Web" [25] involves more attributes than the definition of adaptive applications. It includes all rules of the latter plus additional rules related to conceptual space, these are:

$$\forall x \in X_{AASW}, V_{ontosubject}(x) = 1$$
$$\forall x \in X_{AASW}, V_{ontoosubject}(x) = 1$$
$$\forall x \in X_{AASW}, V_{ontocondition}(x) = 1$$
$$\forall x \in X_{AASW}, V_{ontoaction}(x) = 1$$

5 Influence of Adaptive Applications on the Field of IT-based Service Management

An adaptive behavior aids in automatically selecting the most appropriate services from among the set of available ones. Such a behavior can (partly) take over this selection process based on analysis and benchmarking of these services. A reliable adaptive selection required a standardized unified description of services, criteria, and decision making models.

5.1 Informal Description

An adaptive application which would overtake the task of automatic selection (or recommendation) of services is specified according to the definition above to have the following:

Has content as adaptation object represented by the set of available services. Here content itself is being adapted rather than presentation, navigation, or user. Relevant services are returned while others are filtered out. However, presentation of the result list can be adapted as well to reflect different relevance degrees when applicable.

Adaptation subject for this application is mainly the user represented by the profile of the enterprise unit that wishes to use the services. Characteristics and preferences of this unit are taken into account as criteria for selecting the relevant services such as preferred cost, minimal return on investment, favorite scalability degree… Other adaptation subjects that are independent from the using unit can be involved as well originating from the overall goal of the enterprise, or from an external parameter such as the value of a certain market share which is neither related to service provider nor to using unit (context). Selection of another enterprise unit can also affect the decision forming thereby a neighborhood adaptation subject.

Adaptation model where the "when/where" and "how" to adapt are specified are represented here by the decision making model. Rules for selecting relevant services specify in their "if" part when the adaptation is to take place. For example, the "if" part of such a rule could be matching the user to a content item: "If department *a* prefers low latency and service has latency of 2 s…". It can match the user to another user, and in turn to a content item utilizing a ternary relation: "If department *b* is close in function to department *a*, and department *b* had chosen service *x*…". The "then" part defines how adaptation is to occur, in most cases the application would not have the privilege to take the action of swapping a service with another on its own, but would rather produce a recommendation to do so. This recommendation can take the form of an ordered (and weighted) results list of most relevant candidates. Instead of changing order however, setting different emphasis on different options can be accomplished by using different zoom levels or dimming irrelevant services.

All the adaptation parameters and process mentioned above are on the first order. Second order adaptation takes place when the decision making model is adapted by a supervisor according to a certain subject. Usually this process is done manually therefore the application is "adaptable" on the second order. In the special case where satisfaction of users with the selected services is sensed to be too low and leads therefore to automatically recommending changes in the decision making model, second-order *adaptivity* behavior has taken place.

In order for the application to be able to effectively compare services from different providers these have to be described in a unified manner. This fact leads to the recommendation of having a conceptual space on content items (services) that aids to classify them and allows automatic matchmaking against subject characteristics. At least a single-faceted hierarchy that classifies services based on their functional properties is required. Even better, a multi-faceted hierarchy can have an additional facet for non-functional properties classification; such properties include performance, security, reliability, and so on. An even better implementation of such a conceptual space is an ontology which adds to the advantages of a multi-faceted hierarchy the ability to define a much richer set of relations upon the services such as "variant of", "similar to", or "composed of". An ontology forms as well a common ground for different providers and consumers to share the specification of their services and requirements respectively.

The application, having to "know" preferences of the using enterprise unit, the decision making model of selection, and the list of providers and their services, is expected to deal properly with such sensitive information. Legal privacy issues are not the main concern in this case as opposed to conventional adaptive applications which store personal information of external users. However, confidentiality of stored information and the risk of it being exposed to service providers or to competition is a matter not to be underestimated for such an application. Moreover, transparency of the adaptation decisions must be stressed. That means that when requested, the application should be able to explain its decisions of filtering out some services while recommending others in a human-understandable language. This feature (scrutability [24]) gets more difficult to satisfy the more complex conceptual models and decision model get.

5.2 Formal Specification

The above-mentioned attributes of the example application x can be expressed according to the formal definition in a set of statements as follows:

- models user as adaptation subject: $V_{suser}(x) = 1$
- models goal as adaptation subject (optional): $V_{sgoal}(x) = 1$
- models neighborhood as adaptation subject (optional): $V_{sneighborhood}(x) = 1$
- models context as adaptation subject (optional): $V_{scontext}(x) = 1$
- models content as adaptation object: $V_{socontext}(x) = 1$
- models presentation as adaptation object (optional): $V_{opresentation}(x) = 1$
- models adaptation condition by matching user to content: $V_{conditionU-C}(x) = 1$
- models adaptation condition by matching user to neighborhood to content: $V_{conditionU-N-C}(x) = 1$
- models adaptation action of inserting/removing result list items: $V_{ainsertremove}(x) = 1$
- models adaptation action of ordering, dimming, zooming result list items (optional):

$$V_{aorder}(x) = 1 \qquad V_{a\dim}(x) = 1 \qquad V_{azoom}(x) = 1$$

- is adaptive on the first order: $V_{1\text{st } adaptive}(x) = 1$
- is adaptable on the second order: $V_{2\text{nd } adaptable}(x) = 1$
- has hierarchy on object: $V_{hierobject}(x) = 1$
- has multi-faceted hierarchy or ontology on object (optional):

$$V_{mhierobject}(x) = 1 \qquad V_{ontoobject}(x) = 1$$

- respects privacy and scrutability:

$$V_{privacy}(x) = 1 \qquad V_{scrutability}(x) = 1$$

From this set of attributes we can conclude that the application belongs to the lower approximation of the set "adaptive applications": $x \in \underline{A}X$, optionally to the lower approximation of the set "collaborative filtering applications": $x \in \underline{A}X_{CF}$, and optionally to the upper approximation of the set "adaptive applications on the semantic web": $x \in \overline{A}X_{AASW}$.

6 Conclusion

In this contribution we have defined adaptive applications as a rough set based on a set of attributes. This definition aids researchers and practitioners analyze and distinguish the different classes of applications in addition to discovering new possibilities for enhancing the adaptiveness in their applications and models.

Later, an example of an adaptive recommendation system for service recommendation is explained. Attributes of this system are discussed based on the set of attributes used for the definition above, and membership of this application in different sets is formally determined. Such a compliance study can be simply carried out on different existing or future applications in order to determine the class of adaptive applications they belong to, and hence the common characteristics of class members, best practices to comply with, and common pitfalls to avoid.

References

1. Pawlak, Z.: Rough sets. Int. J. Comput. Inform. Sci. **11**(5), 341–356 (1982)
2. Lenz, C.: Empfaengerorientierte Unternehmenskommunikation Einsatz der Internet-Technologie am Beispiel der Umweltberichterstattung. PhD thesis, Eul, Lohmar; Koeln (2003)
3. Goy, A., Ardissono, L., Petrone, G.: Personalization in e-commerce applications. In: Brusilovsky, P., Kobsa, A., Nejdl, W. (eds.) The Adaptive Web, pp. 485–520. Springer-Verlag, Berlin (2007)

4. De Bra, P., Houben, G.-J., Wu, H.: AHAM: a Dexter-based Reference Model for Adaptive Hypermedia, pp. 147–156. ACM, Darmstadt, Germany (1999)
5. Brusilovsky, P.: Efficient Techniques for Adaptive Hypermedia. Intelligent Hypertext: Advanced Techniques for the World Wide Web, pp. 12–30. Springer-Verlag, London (1997)
6. Oppermann, R. (ed.): Adaptive user support: ergonomic design of manually and automatically adaptable software. L. Erlbaum Associates Inc., Hillsdale (1994)
7. Halasz, F., Schwartz, M.: The Dexter hypertext reference model. Commun. ACM **37**, 30–39 (1994)
8. Mooij, A.D., Cristea, A.I.: LAOS: layered WWW AHS authoring model and their corresponding algebraic operators. In: WWW03 The Twelfth International World Wide Web Conference, Alternate Track on Education, Budapest, Hungary (2003)
9. Lin, Y.l., Brusilovsky, P.: Towards open corpus adaptive hypermedia: A study of novelty detection approaches. In: Konstan, J., Conejo, R., Marzo, J., Oliver, N. (eds.) User Modeling, Adaption and Personalization, Lecture Notes in Computer Science, vol. 6787, pp. 353–358. Springer Berlin (2011)
10. Memari, A., Heyen, C., Marx Gómez, J.: A component-based framework for adaptive applications. In: Modeling of Business Information Systems (MoBIS), Dresden, Germany (2010)
11. Memari, A., Wagner vom Berg, B., Marx Gmez, J.: An agent-based framework for adaptive sustainable transportation. In: 20th IEEE International Workshops on Enabling Technologies: Infrastructures for Collaborative Enterprises, WETICE 2011, Paris, 27–29 June 2011, Proceedings of the IEEE Xplore, Paris (June 2011)
12. Sama, M., Elbaum, S., Raimondi, F., Rosenblum, D.S., Wang, Z.: Context-Aware adaptive applications: Fault patterns and their automated identification. IEEE Trans. Software Eng. **36**(5), 644–661 (2010) 99(PrePrints)
13. Niu, W., Kay, J.: Pervasive personalisation of location information: Personalised context ontology. In: Nejdl, W., Kay, J., Pu, P., Herder, E. (eds.) Adaptive Hypermedia and Adaptive Web-Based Systems. Lecture Notes in Computer Science, vol. 5149, pp. 143–152. Springer Berlin (2008) 10.1007/978-3-540-70987-9 17
14. Brusilovsky, P.: Methods and techniques of adaptive hypermedia. User Model. User-Adap. Inter. **6**(2), 87–129 (1996)
15. De Bra, P., Houben, G., Wu, H.: AHAM: a dexter-based reference model for adaptive hypermedia. In: HYPERTEXT'99 Proceedings of the tenth ACM Conference on Hypertext and hypermedia: returning to our diverse roots: returning to our diverse roots, pp. 147–156. ACM, Darmstadt (1999)
16. Wu, H., De Bra, P., Aerts, A.T.M., Houben, G.: Adaptation control in adaptive hypermedia systems. In: Proceedings of the International Conference on Adaptive Hypermedia and Adaptive Web-Based Systems. AH'00, London, UK, Springer-Verlag (2000) 250-259
17. Knutov, E., De Bra, P., Pechenizkiy, M.: AH 12 years later: a comprehensive survey of adaptive hypermedia methods and techniques. New Rev. Hypermedia Multimedia **15**(1), 5–38 (2009)
18. Brusilovsky, P., Sosnovsky, S., Yudelson, M.: Addictive links: the motivational value of adaptive link annotation. New Rev. Hypermedia Multimedia **15**(1), 97–118 (2009)
19. Archer, P., Smith, K., Perego, A.: Protocol for web description resources (POWDER): description resources. http://www.w3.org/TR/2009/REC-powder-dr-20090901 (Jan 2009)
20. Gruber, T.R.: A translation approach to portable ontology specifications. Knowledge Acquisition **5**(2), 199–220 (1993)
21. Ghali, F., Cristea, A.I.: Social reference model for adaptive web learning. In: Spaniol, M., Li, Q., Klamma, R., Lau, R.W.H. (eds.) Advances in Web Based Learning ICWL 2009, vol. 5686, pp. 162–171. Springer Berlin Heidelberg, Berlin (2009)
22. Ghali, F., Cristea, A.I.: Social reference model for adaptive web learning. In: Spaniol, M., Li, Q., Klamma, R., Lau, R.W.H. (eds.) Advances in Web Based Learning ICWL 2009, vol. 5686, pp. 162–171. Springer Berlin Heidelberg, Berlin, Heidelberg (2009)

23. Kasanoff, B.: Making It Personal: How To Profit From Personalization Without Invading Privacy. 1st edn. Basic Books (November 2001)
24. Kay, J.: Scrutable adaptation: Because we can and must. In: Adaptive Hypermedia and Adaptive Web-Based Systems. Springer Berlin Heidelberg (2006) 11–19
25. Memari, A., Marx Gómez, J.: A model for adaptive applications on the semantic web. In: 3rd International Conference on Information and Communication Technologies: From Theory to Applications, ICTTA'08. (2008)

Chapter 8
Attitude-based Consensus Model for Heterogeneous Multi-criteria Large-Scale Group Decision Making: Application to IT-based Services Management

Iván Palomares and Luis Martínez

Abstract IT-based services management in organizations frequently requires the use of decision making approaches. Several multi-criteria and group decision making models have been proposed for the management of IT-based services in the literature. However, there are some important aspects when a large number of decision makers take part, that have not been considered yet in these organizational contexts, such as: the existence of multiple subgroups of decision makers with different attitudes and/or interests, the necessity of applying a consensus reaching process to make highly accepted collective decisions, and the problem of dealing with heterogeneous contexts, since decision makers from different areas might provide preferences in different information domains. This chapter proposes an attitude-based consensus model for IT-based services management, that deals with heterogeneous information and multiple criteria. An example that illustrates its application to a real-life problem about selecting an IT-based banking service for its improvement is also presented.

Keywords IT-based services · Multi-criteria group decision making · Consensus reaching processes · Heterogeneous information · Attitude

---

I. Palomares (✉) · L. Martínez
Department of Computer Science, University of Jaén,
Campus Las Lagunillas s/n, 23071 Jaén, Spain
e-mail: ivanp@ujaen.es

L. Martínez
e-mail: martin@ujaen.es

M. Mora et al. (eds.), *Engineering and Management of IT-based Service Systems*, 155
Intelligent Systems Reference Library 55, DOI: 10.1007/978-3-642-39928-2_8,

1 Introduction

Nowadays, the management of Information Technology (IT) services (or simply IT-based services) has become an important aspect to guarantee a good operation in many organizational environments [1]. IT-based services management usually considers a number of different processes, such as: the design of new IT services to be introduced in the organization [2], the benchmark and/or analysis of already existing ones to improve or replace them [3, 4], and the choice of the most appropriate IT services' external provider [5], amongst others. Some of these processes might require the use of decision making models, since different alternatives (i.e. IT-based services) are present and the organization must evaluate them and make a decision about determining the most suitable one/s.

The evaluation of IT-based services under a decision making framework usually needs to consider several criteria of different nature (e.g. cost, customer satisfaction, ease of use) [4, 5], in which case a Multi-Criteria Decision Making (MCDM) problem is defined. In a MCDM problem, each alternative must be evaluated according to a finite set of comprehensive criteria, which should be satisfied to as much extent as possible [6–9]. Additionally, the participation of several members belonging to different areas and departments of the organization in processes to manage IT services, make it necessary to deal with Group Decision Making (GDM) problems [10]. In a GDM problem, a group of decision makers or experts must provide their opinions about a set of alternatives in order to make a common decision [11]. Multi-criteria Group Decision Making (MCGDM) problems assume the existence of both multiple criteria and several decision makers.

In MCGDM applied to the management of IT-based services, it is nowadays common that an increasingly larger number of experts belonging to diverse areas of expertise take part in the problem, because IT-based services are used across organizations. Since such experts might present different profiles according to their background, each one may prefer to express his/her preferences about each IT service by using different information domains (e.g. numerical, interval-valued, linguistic). Criteria may also have either a quantitative or a qualitative nature, therefore different information domains might be required to evaluate IT services according to each criterion. In both cases, the MCGDM problem must be defined in an heterogeneous framework, hence an approach to deal with heterogeneous information becomes necessary [12].

The increasing number of decision makers involved in real GDM problems in organizational contexts may also imply the existence of several subgroups of interest with conflicting opinions and attitudes, which leads to possible situations of disagreement amongst them. In such cases, consensus reaching processes (CRPs) to achieve agreed group decisions would be highly appreciated [13], since such decisions may have a more beneficial effect on the organization. In addition, considering the attitude of decision makers towards agreement, i.e. the capacity they present to modify their own preferences to achieve an agreement, would help improving and optimizing CRPs [14].

Despite several MCDM and GDM models have been proposed in the literature to conduct IT services management processes [2–5, 10], they present some noticeable drawbacks: (i) information must be expressed in just one information domain imposed by the model, thus preventing decision makers from using their most preferred domain, (ii) no CRPs are conducted to achieve a high level of collective agreement before making a decision, and (iii) the attitude of decision makers towards agreement is not considered. Therefore, a model that takes these aspects into account becomes necessary in this type of real-life problems.

This chapter proposes an approach to solve large-scale MCGDM problems in organizational environments, that addresses the weaknesses stated above, thus providing some advantages over previous decision approaches proposed in this application context. To do so, we use an attitude-based consensus model [14, 15], which is able to manage heterogeneous information effectively and has been extended to manage multiple criteria.

This chapter is set out as follows: in Sect. 2, some preliminaries related to decision making applied to IT-based services management, heterogeneous MCGDM problems and CRPs are reviewed. Section 3 describes the proposed consensus model, that integrates attitudes in CRPs and manages heterogeneous information. Section 4 shows an application in a real-life problem about the selection of an IT-based banking service. Finally, some concluding remarks are given in Sect. 5.

2 Preliminaries

In this section, we review the management of IT-based services under the point of view of decision making, and revise the formalization of heterogeneous MCGDM problems and some basic notions about CRPs.

2.1 Decision Approaches for IT-based Services Management

IT-based services are becoming a key integrated component of business operations in many organizational contexts nowadays [1]. The increasing presence of IT resources (e.g. the Internet, multi-media and mobile technologies, etc.) as part of the daily operation in most organizations and companies, have raised a high demand for the management of IT-based services, aimed to satisfy business needs and improve social and economic value in the organization [2]. Frequently, processes to manage IT-based services can be viewed as problems that require the use of decision making models, in which different options or alternatives (normally IT services) are considered and evaluated, and the best one/s must be chosen to put it/them in practice [2, 4, 5, 10].

Some examples of proposals based on decision making models to deal with the problem of managing IT-based services can be found in the literature. In [2], an approach based on the analytic hierarchy process (AHP) [16] for MCDM is proposed for the selection of the IT-based service design that best aligns business needs. Another decision making procedure based on multiple criteria is applied in [4], where Multi-Attribute Utility Theory (MAUT) [17] is used to support decisions in IT service portfolio management, dealing with questions such as the ranking of different services according to a set of conflicting criteria under uncertainty. The selection of a provider for acquiring IT services externally is another frequent problem in real companies, which is addressed in [5] by applying fuzzy logic-based approaches in diverse MCDM methods, including AHP and TOPSIS (Technique for Order Preference by Similarity to Ideal Solution) [18], amongst others. In [10], a context-independent GDM model is applied to the problem of IT service continuity management, providing decision makers with a visual representation (based on Hasse diagrams) of their preferences with respect to those provided by the rest of the group.

Regarding MCDM problems, in [19] the most frequent tangible and intangible criteria used for selecting IT services are provided. Table 1 summarizes them in decreasing priority order. Due to the different nature (quantitative or qualitative) that these criteria may present, dealing with several of them makes it apparent the need for managing heterogeneous information [12]. Some of these criteria will be considered in the illustrative example shown in Sect. 4.

2.2 Characterization of Heterogeneous MCGDM Problems

GDM problems are characterized by the participation of two or more experts in a decision problem, where a set of alternatives or possible solutions to the problem are presented [11, 20].

Formally, the main elements found in a GDM problem are:

Table 1 10 most common tangible and intangible criteria for the selection of IT-based services

Tangible	Intangible
Financial cost or loss	Customer experience
Repair time	Faster decisions
Return of Investment (ROI)	Competitive advantage
Resource utilization	Ease of implementation
Revenue	Support availability
Profit margin	Regulation compliance
Internal return rate or payback period	Company's image
Number of people affected by IT fault	Complexity of work-flow for supporting business process
Market share	Service scalability
Share value	Environmental or social contribution

– A set $X = \{x_1, \ldots, x_n\}$, $(n \geq 2)$ of *alternatives* to be chosen as possible solutions to the problem.
– A set $E = \{e_1, \ldots, e_m\}$, $(m \geq 2)$ of *decision makers* or *experts*, who express their judgements or opinions on the alternatives in X.

Each expert e_i, $i \in \{1, \ldots, m\}$, provides his/her opinions over a finite set of alternatives in X by means of a preference structure [21], for instance a preference relation (as will be considered in this proposal), P_i, expressed in an information domain D_i, $\mu_{P_i} : X \times X \rightarrow D_i$,

$$P_i = \begin{pmatrix} - & \cdots & p_i^{1n} \\ \vdots & \ddots & \vdots \\ p_i^{n1} & \cdots & - \end{pmatrix}$$

where the assessment $p_i^{lk} = \mu_{P_i}(x_l, x_k)$ indicates e_i's degree of preference of alternative x_l over x_k, $l, k \in \{1, \ldots, n\}, l \neq k$ [22].

As pointed out in the introduction, in a GDM problem defined in an organizational context with a large number of experts, they might belong to different areas of expertise, therefore they might have different levels of knowledge about such a problem. In these situations, each expert e_i could prefer to express his/her preferences by using an specific information domain D_i, i.e. the GDM problem is defined in an heterogeneous framework. In this proposal, the following information domains $D_i \in \{numerical, interval-valued, linguistic\}$ are considered to assess preferences [12]:

– *Numerical*: Assessments p_i^{lk} are represented as values in [0, 1]. Preference relations based on this information domain are better known as fuzzy preference relations [23, 24].
– *Interval-valued*: Assessments p_i^{lk} are represented as intervals, $I([0, 1])$ [12].
– *Linguistic*: Assessments p_i^{lk} are represented as linguistic terms $s_u \in S$, where $S = \{s_0, \ldots, s_g\}$ is a linguistic term set, and $u \in \{0, \ldots, g\}$ [25].

Besides the need for managing heterogeneous information, GDM problems applied to IT services management usually require that experts evaluate each IT service according to more than one criterion. In such a case, we have an heterogeneous MCGDM problem [6–9] where, given a set of criteria $Z = \{z_1, \ldots, z_q\}$, $q \geq 2$, with quantitative and/or qualitative nature, each expert must provide a preference relation $P_{ij} = (p_{ij}^{lk})^{n \times n}$, $p_{ij}^{lk} \in D_{ij}$, for each criterion $z_j \in Z$.

2.3 Consensus Reaching Processes

Traditional solving processes to find a solution to GDM and MCGDM problems [26], do not guarantee that the decision made would be accepted by the whole group, since some experts might consider that their opinions have not been sufficiently considered [13]. In order to overcome this drawback, CRPs have attained a great attention as part of the decision process. In a CRP, experts discuss and modify their preferences to achieve a high degree of collective agreement in the group before making a decision [13].

The notion of consensus can be interpreted in different ways, ranging from consensus as total agreement, whose accomplishment is usually very costly in practice, to more flexible and feasible approaches that allow considering different degrees of partial agreement to decide about the existence of consensus [27–29]. Amongst such flexible approaches, the so-called notion of *soft consensus* proposed by Kacrpyzk stands out [11, 29]. This approach is based on the concept of fuzzy linguistic majority, according to which consensus exists if *most decision makers participating in a problem agree with the most important alternatives*. The notion of *soft consensus* can be easily reflected in a CRP by using OWA (Ordered Weighted Averaging) operators based on linguistic quantifiers to measure consensus [30].

The process to reach a consensus in GDM problems is an iterative and dynamic discussion process [13], frequently coordinated by a human figure known as moderator, who is responsible for supervising and guiding experts in the overall process, as well as giving them advice to modify their opinions [31]. A basic scheme to conduct CRPs (see Fig. 1) is described below:

(1) *Gather preferences*: Each expert provides moderator a preference structure with his/her opinion on the existing alternatives.
(2) *Determine degree of consensus*: The moderator computes the level of agreement in the group by means of a *consensus measure* [28], usually based on different *similarity measures* and *aggregation operators* [32].

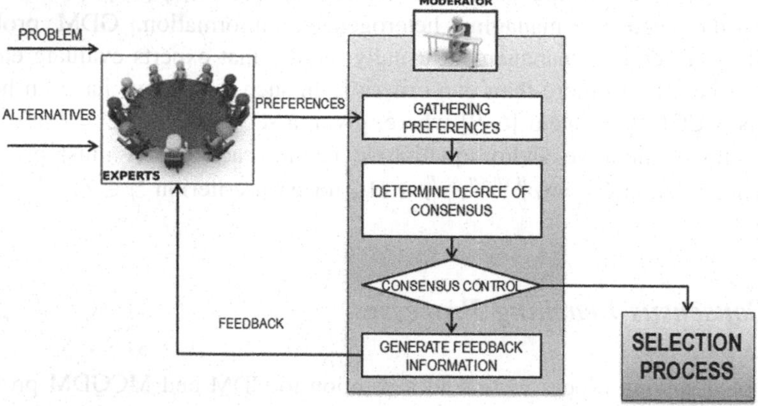

Fig. 1 General consensus reaching process scheme in GDM problems

(3) *Consensus control*: The consensus degree is compared with a threshold level of agreement desired by the group. If such degree is enough, the group moves onto the selection process, otherwise, more discussion rounds are required.
(4) *Generate feedback information*: The moderator identifies furthest preferences from consensus and gives experts some pieces of advice, suggesting them how to modify their opinions and make them closer. Afterwards, a new round of discussion begins.

This basic scheme has been extended to deal with additional features usually present in many real-life decision problems (e.g. managing multiple criteria, heterogeneous information, etc.), as will be shown in the scheme of the consensus model described in the following section.

In the research field of GDM and CRPs, different models have been proposed in the literature [6, 9, 25, 33]. Despite they have been widely applied to solve GDM problems with a few number of experts, most of them do not fit real-life problems with large groups. We propose a consensus model to manage large groups and address all current challenges and limitations.

3 Attitude-based Heterogeneous Consensus Model for Large-Scale MCGDM Problems

As stated in the introduction, some unsolved challenges regarding the application of MCGDM approaches when dealing with large groups of decision makers are: (i) an approach to manage heterogeneous information (experts' preferences expressed in different information domains) becomes necessary, (ii) a CRP should be conducted to prevent possible situations of disagreement amongst some subgroups of decision makers with conflicting interests, and (iii) the CRP could be improved if the group's attitude towards consensus is integrated.

In order to solve such challenges in this type of real-life problems, here it is proposed the use of an attitude-driven consensus model initially presented in [14, 15], and extended here for the resolution of heterogeneous MCGDM problems focused on large groups of decision makers in organizations. The consensus model is characterized by incorporating the following tools:

(a) An approach to integrate the group's attitude towards consensus in the CRP, by means of an aggregation operator so-called Attitude-OWA (Sect. 3.1.1) [14].
(b) An unification-based approach to manage heterogeneous information (Sect. 3.1.2) [12].

Once reviewed these two approaches, a detailed scheme of the phases conducted in the model will be presented, and an illustrative example of its application to a real-life IT-based services management problem will be further described in Sect. 4.

3.1 Approaches to Manage Group's Attitudes and Heterogeneous Information

In the following, the two approaches included in the proposed consensus model to integrate the group's attitude and deal with heterogeneous information will be briefly reviewed.

3.1.1 Incorporating Group's Attitude in CRPs

The concept of attitude towards consensus can be defined as follows:

Definition 1 [14, 15] The group's attitude towards consensus refers to the degree of importance given by decision makers to reach agreement, compared to modifying their own preferences, during a CRP:

- An optimistic attitude means that achieving an agreement is more important for experts than their own preferences, therefore more importance must be given to those positions in the group whose level of agreement is higher.
- A pessimistic attitude means that experts tend to preserve their initial preferences, rather than trying to seek a collective agreement, therefore more importance must be given to positions in the group with lower agreement.

The attitude is integrated across the CRP to determine the degree of agreement in the group (phase (2) in Sect. 2.3). To do so, an aggregation operator, so-called Attitude-OWA, is defined. Such an operator extends OWA aggregation operators [34], which have proved to be appropriate to manage the attitudinal character of aggregation [32], as well as having been successfully applied to a vast array of decision making approaches [35–37]. In order to define Attitude-OWA, two *attitudinal parameters* that must be gathered by the decision group, $\vartheta, \varphi \in [0, 1]$, are introduced:

- ϑ represents the group's attitude, which can be optimistic ($\vartheta > 0.5$), pessimistic ($\vartheta < 0.5$) or indifferent ($\vartheta = 0.5$). It is equivalent to the measure of optimism (*orness*) that characterizes OWA operators [34].
- φ indicates the amount of agreement positions which are given non-null weight in the aggregation process. The higher φ, the more values are considered.

Attitude-OWA operator is then defined as follows:

Definition 2 [14] An *Attitude-OWA operator* of dimension h on a set $A = \{a_1, \ldots, a_h\}$, is an OWA operator based on attitudinal parameters ϑ, φ given by a group of decision makers to indicate their attitude towards consensus,

$$Attitude - OWA_W(A, \vartheta, \varphi) = \sum_{j=1}^{h} w_j b_j \qquad (1)$$

where and A is the set of values to aggregate, and b_j is the jth largest of a_i values.

Attitude-OWA is mainly characterized by its method to compute the weighting vector W, based on the definition of a RIM (Regular Increasing Monotone) linguistic quantifier upon attitudinal parameters [38, 39]. Weights obtained are intended to reflect the specific attitude adopted by decision makers faithfully. The following scheme was proposed in [14] to define Attitude-OWA weights at the beginning of a CRP:

(i) The group determines the values for ϑ, φ, based on their needs and/or the nature of the GDM problem.
(ii) A RIM quantifier with linear membership funcion $Q(r)$, $r \in [0,1]$:

$$Q(r) = \begin{cases} 0 & \text{if } r \leq \alpha, \\ \frac{r-\alpha}{\beta-\alpha} & \text{if } \alpha < r \leq \beta, \\ 1 & \text{if } r > \beta. \end{cases} \quad (2)$$

is defined upon ϑ, φ, by computing $\alpha, \beta \in [0,1]$ as follows:

$$\alpha = 1 - \vartheta - \frac{\varphi}{2} \quad (3)$$

$$\beta = \alpha + \varphi \quad (4)$$

(iii) The following method proposed by Yager in [30], is applied to compute weights w_i:

$$w_i = Q\left(\frac{i}{h}\right) - Q\left(\frac{i-1}{h}\right), i = 1, \ldots, h \quad (5)$$

The effects of integrating different group attitudes in the CRP and their appropriateness in GDM problems with a large number of experts are further discussed and illustrated in [14]. The complete process to define an Attitude-OWA operator upon a group's attitude is shown in Fig. 2.

3.1.2 Management of Heterogeneous Information

In order to deal with assessments p_{ij}^{lk} provided by experts in different domains, we consider the method proposed by Herrera et al. [12] that unifies heterogeneous information into a single domain, consisting in fuzzy sets $F(S_T)$ in a common linguistic term set $S_T = \{s_0, \ldots, s_g\}$ [40]. S_T is chosen according to the guidelines proposed in [12]. To do so, a transformation function is defined for each

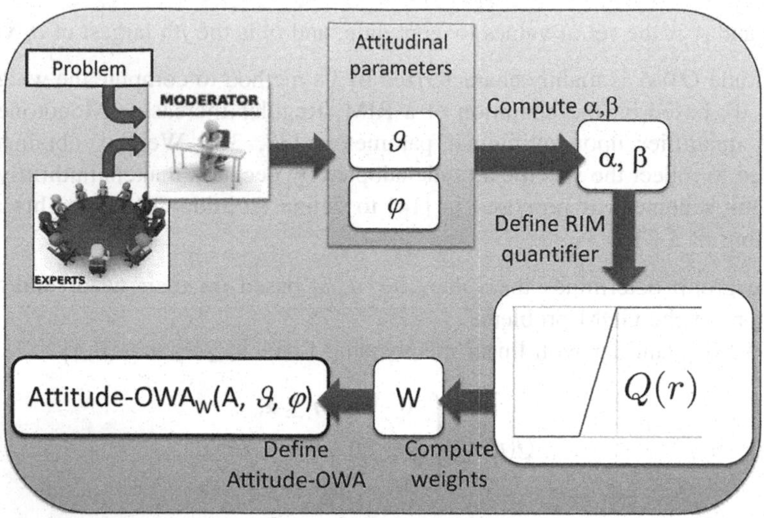

Fig. 2 Process to determine the Attitude-OWA operator used to measure consensus based on the group's attitudinal parameters ϑ and φ

information domain (numerical, interval-valued and linguistic) previously introduced in Sect. 2.2.

Definition 3 [12] Let $v \in [0, 1]$ be a numerical value. The function τ_{NS_T}:$[0, 1] \rightarrow F(S_T)$ transforms it into a fuzzy set in S_T.

$$\tau_{NS_T}(v) = \{(s_0, \gamma_0), \ldots, (s_g, \gamma_g)\}\ s_u \in S_T, \gamma_u \in [0, 1]$$

$$\gamma_u = \mu_{s_u}(v) \begin{cases} 0 & \text{if } v \notin support(\mu_{s_u}(x)), \\ \frac{v-a_u}{b_u-a_u} & \text{if } a_u \leq v \leq b_u, \\ 1 & \text{if } b_u \leq v \leq d_u, \\ \frac{c_u-v}{c_u-d_u} & \text{if } d_u \leq v \leq c_u, \end{cases}$$

being $\mu_{s_u}(\cdot)$ the membership function of linguistic term $s_u \in S_T$, $u \in \{0, \ldots, g\}$, which is represented by a parametric function (a_u, b_u, d_u, c_u).

Definition 4 [12] Let $I = [d, e] \subseteq [0, 1]$ be an interval-valued in [0,1]. The function $\tau_{IS_T} : I \rightarrow F(S_T)$ transforms it into a fuzzy set in S_T.

$$\tau_{IS_T}(I) = \{(s_u, \gamma_u)/u \in \{0, \ldots, g\}\}$$

$$\gamma_u = max_y min\{\mu_I(y), \mu_{s_u}(y)\}$$

where $\mu_I(\cdot)$ and $\mu_{s_u}(\cdot)$ are the membership functions of the fuzzy sets associated with the interval-valued I and linguistic term s_u, respectively.

Definition 5 [12] Let $S = \{l_0, \ldots, l_p\}$ and $S_T = \{s_0, \ldots, s_g\}$ be two linguistic term sets, such that $g \geq p$. A linguistic transformation function $\tau_{SS_T} : S \rightarrow F(S_T)$ transforms a linguistic term l_i into a fuzzy set in S_T.

$$\tau_{SS_T}(l_i) = \{(s_u, \gamma_u^i)/u \in \{0, \ldots, g\}\} \forall l_i \in S$$

$$\gamma_u^i = max_y min\{\mu_{l_i}(y), \mu_{s_u}(y)\}$$

where $i \in \{0, \ldots, p\}$. $\mu_{l_i}(\cdot)$ and $\mu_{s_u}(\cdot)$ are the membership functions of the fuzzy sets associated with the terms l_i and s_u, respectively.

Regarding heterogeneous MCGDM problems as defined in 2.2, once applied these transformation functions on assessments p_{ij}^{lk} to unify them, and assuming that each fuzzy set is represented by its membership degrees $p_{ij}^{lk} = (\gamma_{ij0}^{lk}, \ldots, \gamma_{ijg}^{lk})$, with $i \in \{1, \ldots, m\}, j \in \{1, \ldots, z\}, l, k \in \{1, \ldots, n\}$, each decision maker's preference relation P_{ij} is represented as follows:

$$P_{ij} = \begin{pmatrix} - & \cdots & (\gamma_{ij0}^{1n}, \ldots, \gamma_{ijg}^{1n}) \\ \vdots & \ddots & \vdots \\ (\gamma_{ij0}^{n1}, \ldots, \gamma_{ijg}^{n1}) & \cdots & - \end{pmatrix}$$

3.2 *Attitude-Driven Consensus Model*

Once it has been clarified the concept of group's attitude towards consensus and revised the Attitude-OWA operator to integrate such an attitude in the CRP, as well as the approach to manage heterogeneous information, an heterogeneous attitude-driven consensus model is introduced. This model provides the necessary tools to easily automate all the human moderator's tasks during the CRP, by implementing it into a *Consensus Support System* [15].

Besides the two approaches previously revised, some features of the proposed consensus model, which extends the one presented in [14, 15] to deal with MCGDM problems, are:

(i) Kacpryzk's notion of *soft consensus* and fuzzy majority [11, 29] are implicitly considered in the computation of the agreement level, by using Attitude-OWA operator.
(ii) In order to deal with *multiple criteria* for evaluating IT services, an aggregation step is conducted on each expert's preference relations P_{ij} to combine them [6–8].

Some initial parameters must be fixed by the decision group before beginning the CRP:

- A *consensus threshold* $\mu \in [0, 1]$, i.e. the minimum level of agreement desired by the group, which determines whether consensus achieved is enough or not to conclude the discussion process. The larger its value, the higher the level of agreement that must be achieved.
- A maximum number of discussion rounds allowed, *Maxrounds* $\in \mathbb{N}$, after which the CRP should finish if μ is not exceeded.

Figure 3 shows the phases conducted in the model, which are described in detail below:

(A) Determining group's attitude

This phase is carried out at the beginning of the CRP. The group's attitude towards consensus and its corresponding Attitude-OWA operator are determined, by assigning a value to attitudinal parameters ϑ and φ, based on the context and characteristics of the decision problem to solve, as well as the experts' individual concerns (see Fig. 2).

(B) Gathering preferences

Each expert $e_i \in E$ provides his/her preferences on alternatives in X to the moderator, by means of q preference relations $P_{ij} = (p_{ij}^{lk})^{n \times n}$, one for each criterion $z_j \in Z$. $p_{ij}^{lk} \in D_{ij}$ represents e_i's assessment on the pair of alternatives (x_l, x_k), according to criterion z_j, expressed in an information domain $D_{ij} \in \{numerical, interval - valued, linguistic\}$.

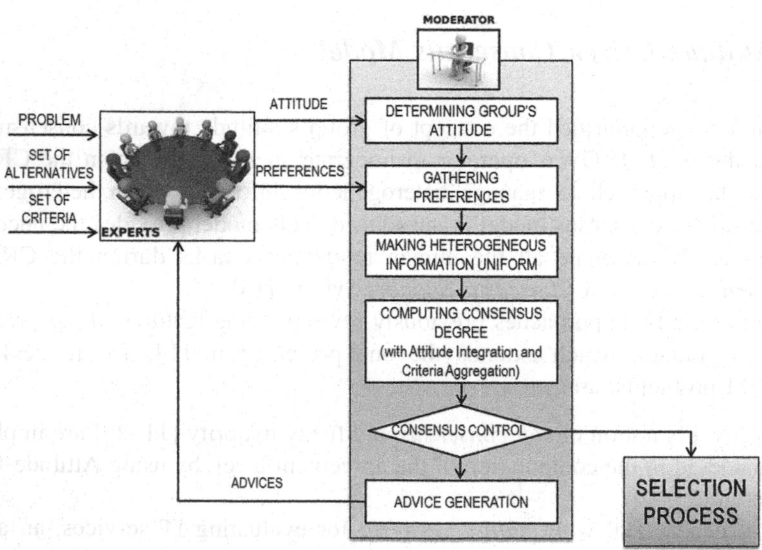

Fig. 3 Attitude-based consensus model scheme

(C) Making Heterogeneous Information Uniform

Our interest focuses on solving MCGDM problems defined in heterogeneous frameworks, where the information about preferences provided by experts can be numerical, interval-valued or linguistic. We consider the method proposed in [12], that unifies information expressed in different domains, p_{ij}^{lk}, into fuzzy sets $F(S_T)$ in a common linguistic term set $S_T = \{s_0, \ldots, s_g\}$, as explained in Sect. 3.1.2.

(D) Computing consensus degree

The moderator computes the level of agreement between experts, by means of the following steps (see Fig. 4):

1. For each assessment p_{ij}^{lk}, $l \neq k$, a *central value* $cv_{ij}^{lk} \in [0, g]$ is computed as follows

$$cv_{ij}^{lk} = \frac{\sum_{u=0}^{g} index(s_u) \cdot \gamma_{iju}^{lk}}{\sum_{u=0}^{g} \gamma_{iju}^{lk}} \tag{6}$$

being $p_{ij}^{lk} = (\gamma_{ij0}^{lk}, \ldots, \gamma_{ijg}^{lk})$ and $index(s_u) = u$.

2. Each expert has associated q preference relations $P_{ij} = (cv_{ij}^{lk})^{n \times n}$, one for each criterion $z_j \in Z$. However, before computing the agreement level, it is necessary to combine such preferences for each e_i, to obtain an overall one, $P_i = (cv_i^{lk})^{n \times n}$, over alternatives in X. Therefore, criteria are aggregated to

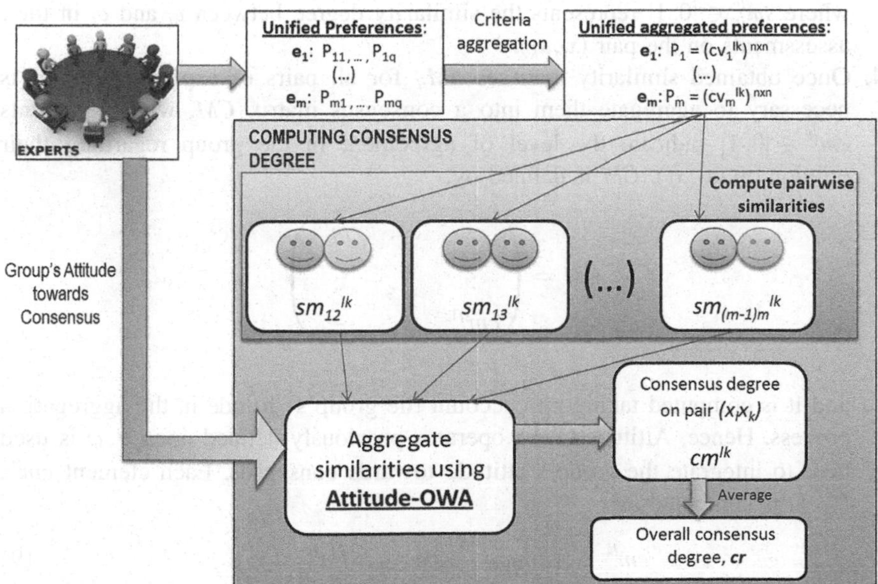

Fig. 4 Procedure to compute consensus degree based on group's attitude

obtain an overall assessment cv_i^{lk} for each e_i and pair (x_l, x_k), by using a weighted averaging operator:

$$cv_i^{lk} = \frac{\sum_{j=1}^{q} \omega_j cv_{ij}^{lk}}{\sum_{j=1}^{q} \omega_j} \tag{7}$$

where $\omega_j \in [0, 1]$, $j = 1, \ldots, q$ are importance weights assigned to criteria z_j, whose values are fixed a priori by the decision group.

3. The computation of consensus degree consists in computing similarity values for each different pair of experts in the group, and aggregating them. Consequently, for each pair e_i, e_t, $i, t \in \{1, \ldots, m\}$, $(i<t)$, a similarity matrix SM_{it}, defined by

$$SM_{it} = \begin{pmatrix} - & \ldots & sm_{it}^{1n} \\ \vdots & \ddots & \vdots \\ sm_{it}^{n1} & \ldots & - \end{pmatrix}$$

must be computed as follows [41]:

$$sm_{it}^{lk} = 1 - \left| \frac{cv_i^{lk} - cv_t^{lk}}{g} \right| \tag{8}$$

where $sm_{it}^{lk} \in [0, 1]$ represents the similarity degree between e_i and e_t in their assessments on the pair (x_l, x_k).

4. Once obtained similarity matrices SM_{it} for all pairs of experts (e_i, e_t), it is necessary to aggregate them into a consensus matrix CM, whose elements $cm^{lk} \in [0, 1]$ indicate the level of agreement in the group regarding their opinion on (x_l, x_k). CM is defined by,

$$CM = \begin{pmatrix} - & \ldots & cm^{1n} \\ \vdots & \ddots & \vdots \\ cm^{n1} & \ldots & - \end{pmatrix}$$

and it is computed taking into account the group's attitude in the aggregation process. Hence, Attitude-OWA operator previously defined upon ϑ, φ is used here to integrate the group's attitude towards consensus. Each element cm^{lk}, $l \neq k$, is computed as:

$$cm^{lk} = Attitude - OWA_W(SIM^{lk}, \vartheta, \varphi) \tag{9}$$

where $SIM^{lk} = \{sm_{12}^{lk}, \ldots, sm_{1m}^{lk}, sm_{23}^{lk}, \ldots, sm_{2m}^{lk}, \ldots, sm_{(m-1)m}^{lk}\}$ is the set of all pairs of experts' similarities in their opinion on (x_l, x_k). Notice that for m experts, we have $\binom{m}{2}$ different pairs of experts (elements to aggregate), therefore $|SIM^{lk}| = \binom{m}{2} = \frac{m(m-1)}{2}$.

5. Once obtained CM, the consensus degree is determined at three different levels to obtain the level of agreement achieved between experts not only on a given pair of alternatives, but also on each alternative and the overall GDM problem:
   (a) Level of pairs of alternatives (cp^{lk}): Obtained from CM as $cp^{lk} = cm^{lk}$, $l, k \in \{1, \ldots, n\}, l \neq k$.
   (b) Level of alternatives (ca^l): The level of agreement on each alternative $x_l \in X$ is computed as:

$$ca^l = \frac{\sum_{k=1, k \neq l}^{n} cp^{lk}}{n-1} \tag{10}$$

   (c) Level of preference relation (overall consensus degree, cr): It is computed as:

$$cr = \frac{\sum_{l=1}^{n} ca^l}{n} \tag{11}$$

(E) Consensus control

The overall consensus degree cr is compared with the consensus threshold $\mu \in [0, 1]$ defined a priori. If $cr \geq \mu$, then the CRP ends and the group moves onto the selection process; otherwise, more discussion rounds are required. *Maxrounds* $\in \mathbb{N}$ is taken into account to limit the number of discussion rounds conducted in the cases that consensus can not be achieved.

(F) Advice generation

If $cr < \mu$, the moderator advices experts to modify their preferences in order to increase the level of agreement in the following rounds. The following steps are considered in this phase (notice that most computations are directly applied on aggregated central values cv_i^{lk}):

1. *Compute a collective preference and proximity matrices for experts*: A collective preference $P_c = (p_c^{lk})^{n \times n}$, $p_c^{lk} \in [0, g]$, is computed for each pair of alternatives by aggregating experts' preference relations:

$$p_c^{lk} = \nu(cv_1^{lk}, \ldots, cv_m^{lk}) \tag{12}$$

where ν is the aggregation operator considered. Afterwards, a proximity matrix PP_i between each expert's preference relation and P_c is obtained:

$$PP_i = \begin{pmatrix} - & \dots & pp_i^{1n} \\ \vdots & \ddots & \vdots \\ pp_i^{n1} & \dots & - \end{pmatrix}$$

Proximity values $pp_i^{lk} \in [0, 1]$ are obtained for each pair (x_l, x_k) as follows:

$$pp_i^{lk} = 1 - \left| \frac{cv_i^{lk} - p_c^{lk}}{g} \right| \tag{13}$$

Proximity values are used to identify the furthest preferences from the collective opinion, which should be modified by some experts.

2. *Identify preferences to be changed (CC)*: Pairs of alternatives (x_l, x_k) whose consensus degree ca^l and cp^{lk} is lower than cr, are identified:

$$CC = \{(x_l, x_k) | ca^l < cr \wedge cp^{lk} < cr\} \tag{14}$$

Afterwards, the model identifies experts who should change their opinion on each of these pairs, i.e. those experts e_i whose assessments on the pair $(x_l, x_k) \in CC$ are such that cv_i^{lk} is furthest to p_c^{lk}. To do so, an average proximity $\overline{pp}^{lk}$ is calculated, by using an aggregation operator λ, as follows:

$$\overline{pp}^{lk} = \lambda(pp_1^{lk}, \dots, pp_m^{lk}) \tag{15}$$

As a result, experts e_i whose $pp_i^{lk} < \overline{pp}^{lk}$ are advised to modify their assessments p_{ij}^{lk} on the pair (x_l, x_k).

3. *Establish change directions*: Several direction rules are applied to suggest the direction of changes proposed to experts, in order to increase the level of agreement in the following rounds. An approach to generate direction rules was proposed in [25]. However, such an approach implied an excessive number of changes, even in the cases that assessments are very close to the collective opinion, therefore we propose extending it by introducing an acceptability threshold $\varepsilon \geq 0$, which should take a positive value close to zero, in order to allow a margin of acceptability when cv_i^{lk} and p_c^{lk} are close to each other.
   - DIR.1: If $(cv_i^{lk} - p_c^{lk}) < -\varepsilon$, then expert e_i should *increase* some of his/her assessments p_{ij}^{lk} associated to the pair of alternatives (x_l, x_k).
   - DIR.2: If $(cv_i^{lk} - p_c^{lk}) > \varepsilon$, then expert e_i should *decrease* some of his/her assessments p_{ij}^{lk} associated to the pair of alternatives (x_l, x_k).
   - DIR.3: If $-\varepsilon \leq (cv_i^{lk} - p_c^{lk}) \leq \varepsilon$ then expert e_i should not modify his/her assessments p_{ij}^{lk} associated to the pair of alternatives (x_l, x_k).

Remark 1 Since experts manage q assessments p_{ij}^{lk} on (x_l, x_k), one for each z_j (instead of managing the unified aggregated assessment, cv_i^{lk}), they decide the extent to which advices are applied to each p_{ij}^{lk}.

4 Illustrative Example in IT-based Services Management

In this section, the proposed attitude-driven consensus model is used to solve a real-life heterogeneous MCGDM problem about selecting an IT-based service offered by a banking company, based on a large group of employees' opinions. In order to automate the human moderator's tasks, the Web-based Consensus Support System presented in [15] has been used, having incorporated the approach to manage multiple criteria in its implementation.

The problem formulation is as follows: A banking company committee compound by 50 employees, $E = \{e_1, \ldots, e_{50}\}$ must make a common decision about determining the IT service offered by the bank which should be modified for its improvement, as part of customer service policies in the organization. Currently, the company provides customers with four different IT-based banking services, $X = \{x_1, x_2, x_3, x_4\}$ [42]:

- x_1:*ATMs*. ATMs (Automated Teller Machines) offer non-stop cash withdrawal, remittances and inquiry facilities. They are used by customers to transact from a network of interconnected branches anywhere and anytime.
- x_2:*Self-inquiry facility*. The terminals situated in the branch allow customers to inquire and view the transactions in the account.
- x_3:*Telebanking*. A 24-hour telephonic service to make inquiries regarding balances and transactions in the account.
- x_4:*Electronic Banking*. This service provides customers with a Graphical User Interface (GUI) software on a personal computer, to perform financial transactions and accounts, cash transfers, cheque book issuing and other inquiries through the Internet.

Employees can use either one of the following information domains:

- Numerical: $[0,1]$
- Interval-valued: $I([0,1])$
- Linguistic: $S = \{s_0 : \textit{null}\ (n), s_1; \textit{very\_low}\ (vl), s_2 : \textit{low}\ (l), s_3 : \textit{medium}\ (m), s_4 : \textit{high}\ (h), s_5 : \textit{very\_high}\ (vh), s_6 : \textit{perfect}\ (p)\}$.

Regarding unification of information, the term set S defined above is chosen as S_T (see Sect. 3.1.2).

Employees must evaluate the existing IT services according to three criteria related to customers, $Z = \{z_1, z_2, z_3\}$ (see Table 1):

- z_1: Amount of people affected by IT fault.
- z_2: Customer experience and satisfaction.

Table 2 Parameters defined at the beginning of the CRP

Criteria weights	$\omega_1 = \omega_3 = 0.5, \omega_2 = 1$
Attitudinal parameters	$\vartheta = 0.4,\ \varphi = 0.7$
Consensus threshold	$\mu = 0.85$
Maximum number of rounds	$Maxrounds = 10$
Acceptability threshold	$\varepsilon = 0.1$

– z_3: Support availability.

Importance weights for criteria, the group's attitude towards consensus and other CRP parameters are summarized in Table 2.

Once defined the problem, the consensus process begins with the first round, following the phases of the consensus model (see Sect. 3.2, Fig. 3):

1. *Determining group's attitude*: The procedure depicted in Fig. 2 is followed to define an Attitude-OWA operator upon attitudinal parameters $\vartheta = 0.4$ and $\varphi = 0.7$. Equations (3) and (4) are used to compute the RIM quantifier's parameters $\alpha = 1 - 0.4 - 0.35 = 0.25$ and $\beta = 0.25 + 0.7 = 0.95$, respectively. Figure 5 shows the RIM quantifier obtained, whose membership function $Q(r)$ is:

$$Q(r) = \begin{cases} 0 & \text{if } r \leq 0.25, \\ \frac{r-0.25}{0.7} & \text{if } 0.25 < r \leq 0.95, \\ 1 & \text{if } r > 0.95. \end{cases} \tag{16}$$

Equation (5) is then used to compute a weighting vector W of dimension $\binom{50}{2} = 1225$ that defines an $Attitude - OWA_W(SIM^{lk}, 0.4, 0.7)$ operator to be later applied in the phase of computing consensus degree (see Sect. 3.2).

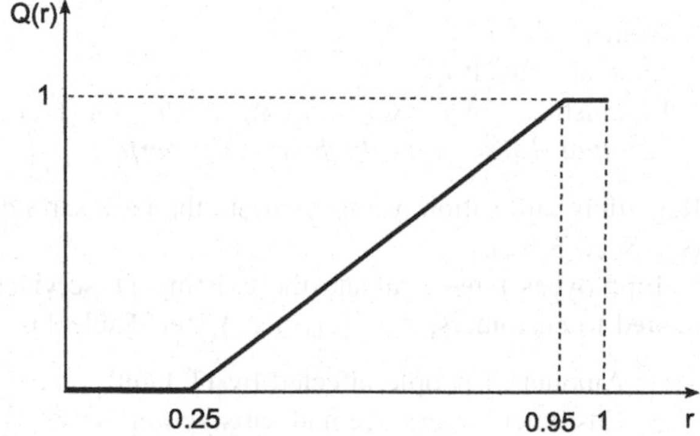

Fig. 5 RIM quantifier defined upon attitudinal parameters $\vartheta = 0.4$, $\varphi = 0.7$

2. *Gathering preferences*: Experts provide their preferences in the form of preference relations, by choosing the information domain they prefer. For instance, an expert e_i provided the following preference relations expressed in a different information domain for each criterion (interval-valued for z_1, linguistic for z_2 and numerical for z_3):

$$P_{i1} = \begin{pmatrix} - & [.6,.8] & [.5,.7] & [.1,.4] \\ [.2,.4] & - & [.7,.9] & [0,.2] \\ [.3,.5] & [.1,.3] & - & [0,0] \\ [.6,.9] & [.8,1] & [1,1] & - \end{pmatrix}, P_{i2} = \begin{pmatrix} - & p & p & h \\ n & - & m & l \\ n & m & - & vl \\ l & h & vh & - \end{pmatrix}$$

$$P_{i3} = \begin{pmatrix} - & .9 & .8, & .5 \\ .1 & - & .4 & 0 \\ .2 & .6 & - & .2 \\ .5 & 1 & .8 & - \end{pmatrix}$$

3. *Making Heterogeneous Information Uniform*: Transformation functions defined in Sect. 3.1.2 are applied on experts' preferences to unify into fuzzy sets in the common linguistic term set $S_T = S$ [12].
4. *Computing consensus degree*: The following steps are conducted to compute the degree of consensus in the group.
   (a) Central values cv_{ij}^{lk} are computed upon unified preferences to ease further computations.
   (b) An overall preference relation P_i, is computed for each e_i by aggregating preference relations P_{ij} for all criteria z_j (see Eq. (7)).
   (c) Similarity matrices are computed for each pair of experts e_i, e_t, $(i<t)$.
   (d) The Attitude-OWA operator defined above is used to aggregate all similarity values for each pair (x_l, x_k), and obtain a consensus matrix CM.
   (e) Consensus degrees at different levels are computed. The global consensus degree cr obtained in the first CRP round is:

$$cr = 0.631$$

5. *Consensus control*: The global consensus degree, $cr = 0.631 < 0.85 = \mu$, therefore consensus amongst employees is not enough and, consequently, the advice generation phase and another CRP round will be required.
6. *Advice generation*: Computations described in Sect. 3.2 are carried out to generate some recommendations for each employee to modify his/her preferences and increase the level of agreement. The arithmetic mean is used here for obtaining the collective preference and average proximity values. Once all

Table 3 Global consensus degree for each round

Round 1	Round 2	Round 3	Round4	Round 5
0.631	0.716	0.766	0.831	**0.868**

experts modify and provide again their preferences, the second round of discussion begins.

In this example, due to the choice of a slightly pessimistic attitude, it was necessary to carry out five rounds of discussion to reach the consensus threshold $\mu = 0.85$ and achieve the level of agreement required. Global consensus degrees obtained in each round are summarized in Table 3.

Once achieved a consensus, the committee proceeds to apply a selection process [43, 44], to determine the IT-based service that must be improved by the company.

The main advantages of using the proposed consensus model to solve MCGDM problems in IT-based services management, can be summarized as follows:

- The attitude-based consensus model provides a flexible scheme to optimize CRPs and achieve highly agreed decisions, based on the presence of subgroups with different interests and the characteristics of each problem.
- The management of multiple criteria in the IT-based services problem, which is addressed by assigning weights to them and aggregating each expert's preferences.
- The approach to unify preferences into a common domain [12], facilitates dealing with heterogeneous frameworks and lets decision makers in the organization express their preferences in the domain they prefer.

5 Concluding Remarks

IT-based services management is a common problem in many organizations nowadays, which frequently requires the use of decision-based approaches to solve it. In this context, although the existence of several decision makers and multiple criteria to evaluate IT services have already been addressed in the literature, some aspects, including the necessity of conducting a consensus reaching process to achieve agreed decisions in large groups, where different subgroups with their own attitudes and interests must cooperate, and the need for managing heterogeneous information, have not been considered yet. In this chapter, we have proposed an heterogeneous consensus model for large-scale group decision making, that integrates the attitude of decision makers. Besides dealing with the heterogeneous information provided by decision makers, such a model has been extended to manage preferences expressed under several criteria. An illustrative example of its application to a real-life problem consisting in selecting an IT-based banking service for its improvement has been also presented.

Acknowledgments This work was partially supported by the Research Project TIN-2009-08286 and ERDF.

References

1. Mora, M., O'Connor, R., Raisinghani, M., Macías-Luévano, J., Gelman, O.: An IT service engineering and management framework (ITS-EMF). Int. J. Serv. Sci. Manag. Eng. Technol. **2**(2), 1–15 (2011)
2. Weist, P.: An AHP-based decision making framework for IT service design. In: MWAIS 2009 proceedings, Paper 11 (2009)
3. Zhu, F., Wymer, W., Chen, I.: IT-based services and service quality in consumer banking. Int. J. Serv. Ind. Manag. **13**(1), 69–90 (2002)
4. Queiroz, M., Moura, A., Sauvé, J., Bartolini, C., Hickey, M.: A framework to support investment decisions using multi-criteria and under uncertainty in IT service portfolio management. In: Proceedings of 2010 IEEE/IFIP network operations and management symposium workshops, pp. 103–110 (2010)
5. Karami, A., Guo, Z.: A fuzzy logic multi-criteria decision framework for selecting IT service providers. In: Proceedings of the 45th Hawaii international conference on system sciences, 2012, pp. 1118–1127 (2012)
6. Parreiras, R., Ekel, P., Martini, J., Palhares, R.: A flexible consensus scheme for multicriteria group decision making under linguistic assessments. Inf. Sci. **180**(7), 1075–1089 (2010)
7. Pokehar, S., Ramachandran, M.: Application of multi-criteria decision making to sustainable energy planning—a review. Sustain. Energy Rev. **8**(4), 365–381 (2004)
8. Zhang, G., Lu, J.: An integrated group decision-making method dealing with fuzzy preferences for alternatives and individual judgements for selection criteria. Group Decis. Negot. **12**(6), 501–505 (2003)
9. Pedrycz, W., Ekel, P., Parreiras, R.: Fuzzy Multicriteria Decision-Making: Models, Methods and Applications. Wiley, Chichester (2011)
10. Van de Walle, B., Rutkowski, A.: A fuzzy decision support system for IT service continuity threat assessment. Decis. Support Syst. **42**(3), 1931–1943 (2006)
11. Kacprzyk, J.: Group decision making with a fuzzy linguistic majority. Fuzzy Sets Syst. **18**(2), 105–118 (1986)
12. Herrera, F., Martínez, L., Sánchez, P.: Managing non-homogeneous information in group decision making. Eur. J. Oper. Res. **166**(1), 115–132 (2005)
13. Saint, S., Lawson, J.R.: Rules for Reaching Consensus: A Modern Approach to Decision Making. Jossey-Bass, San Francisco (1994)
14. Palomares, I., Liu, J., Xu, Y., Martínez, L.: Modelling experts' attitudes in group decision making. Soft. Comput. **16**(10), 1755–1766 (2012)
15. Palomares, I., Rodríguez, R., Martínez, L.: An attitude-driven web consensus support system for heterogeneous group decision making. Expert Syst. Appl. **40**(1), 139–149 (2013)
16. Bryson, N.: Group decision-making and the analytic hierarchy process. exploring the consensus-relevant information content. Comput. Oper. Res. Int. J. **23**(1), 27–35 (1996)
17. Keeney, R., Raiffa, H.: Decisions with Multiple Objectives: Preferences and Value Tradeoffs. Cambridge University Press, Cambridge (1993)
18. Lai, Y., Liu, T., Hwang, C.: TOPSIS for MODM. Eur. J. Oper. Res. **76**(3), 486–500 (1994) (Cited By (since 1996): 116)

19. Moura, J.: Survey and interviews on IT financial management. http://www.bottomlineproject.com/hp/_media/survey_and_interviews_on_it_financial_management.pdf (2008)
20. Butler, C., Rothstein, A.: On Conflict and Consensus: A Handbook on Formal Consensus Decision Making. Food Not Bombs Publishing, Takoma Park (2006)
21. Herrera-Viedma, E., Herrera, F., Chiclana, F.: A consensus model for multiperson decision making with different preference structures. IEEE Trans. Syst. Man Cybern. Part A: Syst. Hum. **32**(3), 394–402 (2002)
22. Fodor, J., Roubens, M.: Fuzzy Preference Modeling and Multicriteria Decision Support. Kluwer, Boston (1995)
23. Orlovsky, S.: Decision-making with a fuzzy preference relation. Fuzzy Sets Syst. **1**(3), 155–167 (1978)
24. Tanino, T.: Fuzzy preference relations in group decision making. In: Kacprzyk, J., Roubens M. (eds.) Non-Conventional Preference Relations in Decision Making, pp. 54–71. Springer, Berlin (1988)
25. Mata, F., Martínez, L., Herrera-Viedma, E.: An adaptive consensus support model for group decision-making problems in a multigranular fuzzy linguistic context. IEEE Trans. Fuzzy Syst. **17**(2), 279–290 (2009)
26. Roubens, M.: Fuzzy sets and decision analysis. Fuzzy Sets Syst. **90**(2), 199–206 (1997)
27. Elzinga, C., Wang, H., Lin, Z., Kumar, Y.: Concordance and consensus. Inf. Sci. **181**(12), 2529–2549 (2011)
28. Herrera-Viedma, E., García-Lapresta, J., Kacprzyk, J., Fedrizzi, M., Nurmi, H., Zadrozny, S. (eds.): Consensual Processes. Studies in Fuzziness and Soft Computing, vol. 267. Springer, Berlin (2011)
29. Kacprzyk, J., Fedrizzi, M.: A "soft" measure of consensus in the setting of partial (fuzzy) preferences. Eur. J. Oper. Res. **34**(1), 316–325 (1988)
30. Yager, R.: Quantifier guided aggregation using OWA operators. Int. J. Intell. Syst. **11**, 49–73 (1996)
31. Martínez, L., Montero, J.: Challenges for improving consensus reaching process in collective decisions. New Math. Nat. Comput. **3**(2), 203–217 (2007)
32. Beliakov, G., Pradera, A., Calvo, T.: Aggregation Functions: A Guide for Practitioners. Springer, Berlin (2007)
33. Xu, J., Wu, Z.: A discrete consensus support model for multiple attribute group decision making. Knowledge-based Syst. **24**(8), 1196–1202 (2011)
34. Yager, R.: On orderer weighted averaging aggregation operators in multi-criteria decision making. IEEE Trans. Syst. Man Cybern. **18**(1), 183–190 (1988)
35. Bordogna, G., Fedrizzi, M., Pasi, G.: A linguistic modeling of consensus in group decision making based on OWA operators. IEEE Trans. Syst. Man Cybern. Part A: Syst. Hum. **27**(1), 126–133 (1997)
36. Yager, R.: Weighted maximum entropy OWA aggregation with applications to decision making under risk. IEEE Trans. Syst. Man Cybern. Part A: Syst. Hum. **39**, 555–564 (2009)
37. Yager, R., Reformat, M., Gumrah, G.: Fuzziness, OWA and linguistic quantifiers for web selection processes. IEEE International conference on fuzzy systems, 2011, pp. 1751–1758
38. Liu, X., Han, S.: Orness and parameterized RIM quantifier aggregation with OWA operators: a summary. Int. J. Approx. Reason. **48**, 77–97 (2008)
39. Zadeh, L.: A computational approach to fuzzy quantifiers in natural languages. Comput. Math. Appl. **9**, 149–184 (1983)
40. Zadeh, L.: The concept of a linguistic variable and its applications to approximate reasoning. Inf. Sci. Part I, II, III **8, 8, 9**, 199–249, 301–357, 43–80 (1975)
41. Herrera-Viedma, E., Martínez, L., Mata, F., Chiclana, F.: A consensus support system model for group decision making problems with multigranular linguistic preference relations. IEEE Trans. Fuzzy Syst. **13**(5), 644–658 (2005)
42. Ghaziri, H.: Information technology in the banking sector: opportunities, threats and strategies. American University of Beirut, Graduate School of Business and Management (1998)

43. Herrera, F., Herrera-Viedma, E., Verdegay, J.: A sequential selection process in group decision making with linguistic assessments. Inf. Sci. **85**(1995), 223–239 (1995)
44. Clemen, R.: Making Hard Decisions: An Introduction to Decision Analysis, 2nd edn. Duxbury Press, Boston (2005)

Chapter 9
Improving Decision-Making for Clinical Research and Health Administration

Alexandra Pomares-Quimbaya, Rafael A. González, Wilson-Ricardo Bohórquez, Oscar Mauricio Muñoz, Olga Milena García and Dario Londoño

Abstract This chapter presents a health decision-support system called DISEArch that allows the identification and analysis of relevant EHR for decision-making. It uses structured and non-structured data, and provides analytical as well as visualization facilities over individual or sets of EHR. DISEArch proves to be useful to empower researchers during analysis processes and to reduce considerably the time required to obtain relevant EHR for a study. The analysis of semantic distance between EHR should also be further developed. As with any information systems project, a conversation needs to be put in place to realize the full potential that IT-based systems offer for people, in this case within the medical domain. It is a mutual learning experience that requires constant translations, frequent prototype discussions, grounding of new IT-based support in current practices and clear identification of existing problems and future opportunities that are opened up in order to enrich the momentum of the project, enlarge the community of early adopters and guaranteeing the continued financial, scientific and administrative support for the project from management stakeholders. Our experience is very

A. Pomares-Quimbaya · R. A. González (✉) · W.-R. Bohórquez · O. Milena García · D. Londoño
Pontificia Universidad Javeriana, Bogotá, Colombia
e-mail: ragonzalez@javeriana.edu.co

A. Pomares-Quimbaya
e-mail: pomares@javeriana.edu.co

W.-R. Bohórquez
e-mail: ricardob@javeriana.edu.co

O. Milena García
e-mail: omgarcia@husi.org.co

D. Londoño
e-mail: dlondono@javeriana.edu.co

O. Mauricio Muñoz
Hospital Universitario San Ignacio, Bogotá, Colombia
e-mail: o.munoz@javeriana.edu.co

M. Mora et al. (eds.), *Engineering and Management of IT-based Service Systems*, Intelligent Systems Reference Library 55, DOI: 10.1007/978-3-642-39928-2_9,

positive and we intend to further pursue this approach and extract lessons learned for similar projects.

Keywords Health-care DSS · IT service system · Data mining

1 Introduction

Decision-making in healthcare is varied and complex: diagnosis, treatment, research and administration are all activities that heavily involve decisions based on evidence and experience, as well as economic and human considerations. IT in healthcare has a had a prominent place within applied information systems research for a long time and has a strong tradition of rigorous and relevant contributions, ranging from the very deeply technical and algorithmic—such as classic and modern uses of artificial intelligence and expert systems [1–3]—to more socio-technically minded interventions—such as the use of Checkland's Soft Systems Methodology mostly within the context of the UK's NHS (e.g. [4, 5]).

Despite evident progress and research dynamics, real impact has not been felt significantly where it matters most. As Hesse and Shneiderman [6] argue, it has probably been a matter of not asking the right questions: rather than focusing on what technology can do, we should be focusing on what people can do (with this technology). This follows a general trend in information systems which pays attention to user-centered, participative design [7, 8]. The popularization of many user-oriented information technologies has shown that user experience and involvement in the design, appropriation and evolution of IT exceed consumer electronics and applications and are indeed morally and pragmatically desirable for information systems development in general.

This creates three conditions for the development of IT to support decision-making in such a sensitive domain: rigor, relevance and user-centered participation. In fact, these requirements are not mutually exclusive but dependent on each other. In this chapter we are guided by those three considerations in presenting ongoing research aimed at contributing to decision-making within a hospital setting with both clinical research and administration as the first stages, but with potentially many other applications in the future. One of the main sources for supporting decision-making in hospitals has been the creation and use of electronic health records (EHR), a rich source of data when properly exploited. Nonetheless, in practice, the use of EHR is more complex due in part to the lack of having considered the three pillars of rigor, relevance and user-centeredness. Often, healthcare professionals are invited (or forced) to adapt to the systems that keep track of patient records, rather than having the system support the professionals in their activities. Of course, striking that balance is not easy, but neglecting it creates problems, such as the tendency to use open text fields to input information that should otherwise be input into structured fields. This narrative approach to recording patient information fits many healthcare practices but unfortunately does not exactly match the logic and technologies offered by current decision-support systems.

One specific approach that has garnered increased attention within the healthcare domain has been data mining [9, 10]. Through the various technologies that can be used for data mining, clinical research and practice can be considerably improved and new patterns can be extracted to help with diagnosis, treatment, and cost-benefit analysis, among others. However, as stated above, when data mining relies on electronic health records, the use of narrative text, makes traditional data mining approaches limited. The objective of this chapter is to describe the process of seeking rigor, relevance and user-centered participation in the design of a support tool for clinical research (at a first stage) and administrative decision-making (at a second stage), using data mining technologies and considering the restrictions imposed by a dataset that is not structured and relies on narrative text.

This research follows a design science research philosophy [11], which has gained increasing support from information systems researchers, given its open goal of providing a framework for research that is both relevant and rigorous. The tension between these often conflicting aspects is dealt with through a design-centered paradigm, where the relevance of solving real-world problems is achieved through a specific design, which in Herbert Simon's tradition is no less than problem-solving itself, i.e. design is problem-solving as it fills the gap between a present situation and a desired one. In addition, design also involves the use of applicable knowledge which becomes embedded in the design process and product, thus being rigorous insofar as this knowledge is applied transparently and systematically. As a result, design science research offers a tripartite framework, relevance-design-rigor, onto which this chapter further elaborates the recommendation that the core design be participative in nature, making it more relevant, as the beneficiaries (be they users, beneficiaries or customers) become co-designers. Specifically, in developing the artifact (as design science research products are dubbed in the Simonian tradition) to support clinical research and administrative decision-making, rigor is supported by a systematic exploration of both literature and technological development, as expressed in patents. Relevance stems from a problem-oriented approach focused on specific needs of a clinical research group at a university hospital. It is in the multi-disciplinary interaction between this research group and an information systems research group that a user-centered participative design has been followed. Together with the participation of other potential users, a prototype has been developed as a proof of concept of the underlying data (text) mining models and algorithms and as a source of medical validation with respect to the quality of the results.

The research approach has followed a problem-initiated process in which an initial relevance cycle (see [12]) has been used to identify requirements, potential users, associated processes (both for clinical research and for administrative purposes), existing technology (the hospital's information system, the underlying databases, the users' capabilities), as well as identifying other hospitals that, given similar technologies, could also benefit from the resulting artifact. A rigor cycle has helped uncover the applicable knowledge by doing a literature and patent analysis of similar problems and applications found scientifically and commercially. Although it could appear as if the rigor cycle follows the relevance cycle, in reality both cycles have moved mostly in parallel. This simultaneous cycling through rigor and relevance is desirable and natural, since by uncovering requirements the researchers have been led to revise and refine the search space within the knowledge base. Conversely, and given the fact that the initial problem was ill-defined and thus still open, as applicable knowledge is found and shared with the medical researchers, this feeds back on the relevance cycle, by making explicit the possibilities offered by existing methods, tools and technologies.

Once the relevance and rigor cycles have offered sufficiently clear requirements and applicable knowledge, they meet inside a design cycle, which iterates between actual design and continued evaluation. This process is akin to classic information systems design, where initial mockups, forms and flow charts are built, refined and evaluated both from a technical point of view and from a potential user's point of view, until these preliminary models can be codified, following a traditional data mining process. Inspired by CRISP-DM [13], understanding the business is followed by understanding the data, after which such data is prepared (to make it amenable for treatment by data mining techniques), models are created to process the data and finally the solution is evaluated and deployed. As can be seen, this data mining process fits naturally within the design science research approach and simply gives it a specific flavor. Evaluation, as stated above, is an iterative process in parallel with the design and refinement of the data mining models and the resulting software prototype. The aim has been to use a test copy of the hospitals EHR database in order to produce the required results and then sharing such results with the complete research team so that evaluation is carried out not only from an information systems (data mining) point of view, but crucially from a medical perspective as well. Final validation of the resulting artifact is still ongoing but is achieved by using the prototype as a proof of concept and analyzing the results so that the desired EHRs are found with precision and recall criteria, or in clinical terms, with enough sensitivity and specificity. The EHRs to be found respond to the requirements, which stem from the following research question.

The main problem and research question has been how to identify patients associated with a specific diagnosis within the set of electronic health records contained in the hospital information system. Though at first this may seem a straightforward question amenable to treatment through simple database queries, the complexity of an actual diagnostic process (for instance, the fact that it is not a point in time, but the result of several events) and the aforementioned nature of the data (unstructured narrative text) make using queries limited and indeed impractical. For example, one may query the system for patients whose EHR contains "diabetes" in a specific field for diagnosis, but very often the diagnosis is not contained in this field in such a clear-cut fashion. Moreover, extending the query to other fields does not solve the problem, because the narrative is nuanced and dynamic (one could find, for instance, "discard diabetes"). Having access to the list of patients with an identified diagnosis is useful given that it is the basis for many of the clinical research studies being carried out by the chronic disease research group involved in this project, but of course of other groups as well. Furthermore, this data is also invaluable for administrative purposes in assessing fitness to treatment protocols, costs and patient distributions.

This chapter is structured as follows. Section 2 presents a systematic review of literature and patents surrounding the use of data mining techniques for patient identification and visualization of electronic health records. Section 3 then presents some of the potential uses of these techniques in supporting decision making both for clinical research and health administration. Section 4 goes on to present the proposal of s system, dubbed DISEarch, to be used for patient identification using health records from a university hospital information system. Some early results of the prototype are discussed in Sect. 5, and then some conclusions and future work are mentioned in Sect. 6.

2 Evolution of Strategies for Healthcare Analytics

Healthcare analytics is increasingly supported by data mining strategies and their associated techniques. In order to uncover the evolution of these strategies, this section discusses the results of a literature review focused on identifying some of the main contributions, institutions and authors centered on this topic. Subsequently, the second part of this section describes some outstanding recent contributions for the improvement of medical data visualization to support healthcare analytics, offering a view of their strengths and opportunities for improvement that have been taken into consideration in designing the proposal brought forward in this chapter.

2.1 Trends in EHR identification and Analysis Technologies

The development of tools, techniques and models for identifying (that is, searching and retrieving EHR according to a specific information need) and analyzing (that is, looking for patterns, classifying or processing EHR for decision making) electronic health records in the last 30 years has been studied through systematic queries in the Scopus database using keywords related to patient, medical or health records, coupled to data preparation, data classification, data and text mining, prioritization, and decision or regression trees. Iteratively, different queries have been used in order to refine the coverage and relevance of the results. In addition, since medical data is particularly sensitive, anonymization, data protection and privacy protection have also been coupled to EHR.

Although the main goal of this research is centered on data mining (and related) strategies for clinical research, most of the work found is aimed at aiding in diagnosis and contributing to effectiveness analysis of particular treatments (or drugs), whether from a medical or an economic point of view. There are few contributions directly related to identifying or analyzing a set of health records for which a specific diagnosis has already been registered for use in clinical research on that specific disease, and given a context in which such diagnosis is hidden in free or narrative text, as described elsewhere in this chapter.

By looking at some of the most cited papers, it can be seen that, for instance in [14], cited 774 times at the time of query, patient records are partitioned according to their being treated with radiation therapy, in order to estimate the effectiveness of the treatment through decision trees. However, as pointed out, these patient records have already been identified as belonging to the interest group, prior to the study. Furthermore, the actual analysis is purely statistical in nature and although most data mining techniques are indeed statistics-based, in building the model proposed in this research, the interest is on techniques supported directly by software. In [15], cited 45 times, a similar contribution is related this time to determining the effectiveness of an antibiotic through regression analysis. In [16], with 265 citations, is more closely related to informatics or computer science techniques, but this time it deviates from the clinical or medical domain into chemistry (drug research), by reporting a data mining algorithm which, through machine learning, can be used to predict the effect of a particular drug, according to its chemical composition. A similar effort is found in [17], except this time the emphasis is on predicting the toxicity of the compound. In [18], cited 176 times, report on algorithms that are used in decision making support for treating bipolar disorder; in this case, given the disorder, certain treatment protocols are designed. For the purposes of the present work, the opposite would be more appropriate; that is, tracing the compliance with a given treatment protocol on a set of patients that have to be identified as having been diagnosed with a given disease—the idea would then be to extract treatment patterns from the identified set and then comparing those patterns to the already established protocol. In [19], cited 156 times, is one of many cases found reporting a cost-benefit analysis, this time related to using prophylactic antiemetic therapy for reducing postoperative nausea and vomiting. Similarly, [20], with 124 citations, evaluates the effectiveness of carrying out a hemocromatosis test in blood donors in order to demonstrate the economic benefit of doing so for a national health system.

One paper potentially more closely related to the purpose of this chapter's contribution is [21], cited 130 times. Their work is aimed at classifying patients into specific groups, according to a set of medical variables. The difference lies in that the starting set of patient records is already known to be diagnosed with unknown primary carcinoma, while our work starts one step behind: identifying patients diagnosed with a specific condition from a complete set of records of diverse patients. Furthermore, Hess et al. only consider quantitative, structured variables, while our work considers also unstructured narrative text. In terms of the sources of these related works, something worth noting is that they are mostly originating in hospitals, medical centers and pharmaceutical labs. Few of the results are explicitly affiliated to collaboration between hospitals and universities or between university hospitals and other departments (especially those related to computer science). Though keyword analysis, several findings are meaningful.

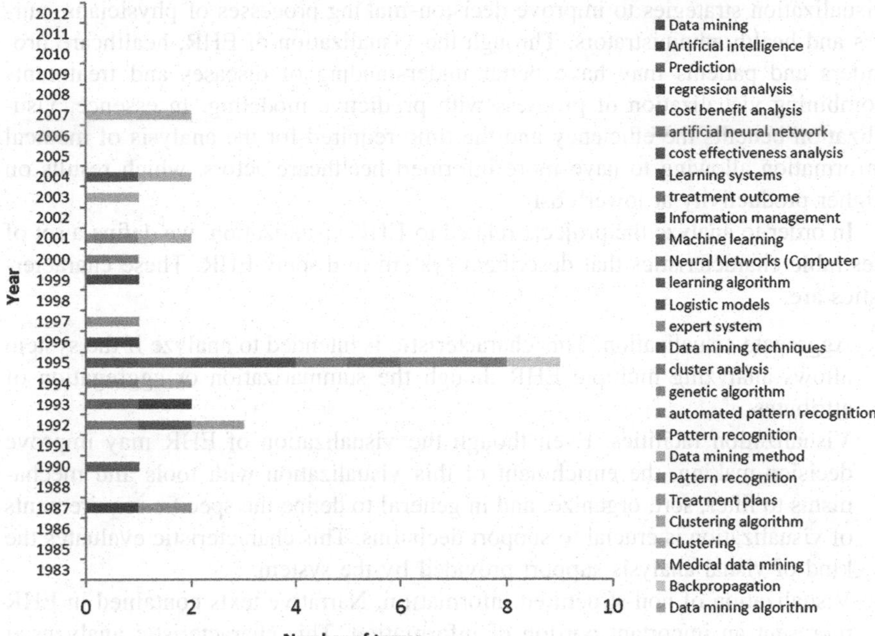

Fig. 1 First appearance of terms as contained in Scopus 1983–2012

Firstly, the historical development of this field of research has been incorporating different techniques as the years go by. Figure 1 shows the moment at which each one has appeared for the first time in the resulting set of records.

In sum, a detailed exploration of relevant works suggests that there have been many contributions to patient identification using data mining techniques and that these have been growing rapidly. However, recalling the introduction of this chapter, there is still work to be done in order to generalize these works into decision-support tools to be used widely for clinical research and health administration. Significantly, it has been found that the majority of works are aimed at sets of data for which a diagnosis, disease or condition has already been determined. In other words, the population for study has already been defined. There is a clear opportunity to contribute works that support the preceding stage; that is, we have found that in practice most clinical research group and health administration must go through a time-consuming, labor-intensive, statistically supported, iteratively queried process of determining patients for study.

2.2 Decision Support Through EHR Visualization

In parallel with the systems and strategies to identify and analyze EHR data, an important group of projects associated with the contribution of this chapter are focused on the visualization of EHR. These projects have advanced in visualization strategies to improve decision-making processes of physicians, nurses and health administrators. Through the visualization of EHR, healthcare providers and patients may have better understanding of diseases and treatments combining visualization of progress with predictive modeling. In essence, visualization benefits the efficiency and the time required for the analysis of medical information allowing to have more informed healthcare actors, which results on higher productivity at lower cost.

In order to analyze the projects related to EHR visualization, we define a set of desirable characteristics that describe a system to display EHR. These characteristics are:

1. Aggregate visualization. This characteristic is intended to analyze if the system allows analyzing multiple EHR though the summarization or aggregation of attributes.
2. Visualization facilities. Even though the visualization of EHR may improve decision-making, the enrichment of this visualization with tools and mechanisms to filter, sort, organize, and in general to define the specific requirements of visualization is crucial to support decisions. This characteristic evaluates the kind of visual analysis support provided by the system.
3. Visualization of non-structured information. Narrative texts contained in EHR represent an important portion of information. This characteristic analyzes if the system allows representing visually this kind of information.
4. User adaptability. Different healthcare actors require different visualizations. This characteristic evaluates whether the visualization system supports the adaptation of what it presents according to the user.
5. Analytic functionalities. Human eyes can detect basic patterns and behaviors; however, sometimes the patterns within a set of EHR are not easily detected by simply visualization. This characteristic evaluates if the system provides analytical functions to identify sequences, patterns and associations.
6. Anonymization. This characteristic evaluates if the system allows to anonymize the EHR before its visualization. Anonymization strategies must be part of Health decision systems for assuring patient confidentiality when his EHR is used during analysis processes.

Table 1 synthesizes the analysis of a representative set of systems that contribute to enhance EHR visualization. These systems include important advances to simplify the visualization of EHR through for analysis purposes. Furthermore, they propose interesting strategies to adapt the content presented according to the type of user.

However, as the table suggests, even though these systems make EHR data easily searchable and accessible, they have limitations for being used for decision-making processes. Particularly, they assume the existence of well-structured EHR that should be visualized, the real situation is that EHR are hardly ever composed only by structured attributes, and the attributes with narrative texts are often the most important. In addition, decision-making processes require the analysis of

Table 1 EHR visualization systems

	Aggregate	Visualization facilities	Non-structured information	Adaptability	Analytic functionalities	Anonymization
PRIMA [1]	+	+	–	–	–	–
LifeLines [22]	+	+	–	–	–	–
RAVEL [23]	–	–	–	+	–	–
Adaptive EHR [24]	–	+	+	+	–	–
Adaptive visual [25]	–	+	+	+	–	–
TimeLine [26]	–	+	–	–	–	–

information from multiple EHR, respecting the confidentiality and providing the relevant view for the decision maker. This capability has not yet been reached completely by the available tools. An extension of their functionalities may provide richer systems to support decision-making.

The aim in this chapter is to propose a model that contributes to identifying patients for a given disease that are then amenable for use in clinical research and health administration. Moreover, a user-centered, participative co-design of the resulting artifacts has not been found as often. As a result, we follow this chapter by setting the case for a multi-disciplinary project carried out in a university hospital.

3 Improving Decision-Making for Clinical Research and Health Administration

As the previous section discusses, electronic health record systems (EHR) have improved the access to patient information by health care providers. They are a rich source of knowledge widely used to improve health care activities such as diagnostic and treatment definition. In addition, they have been also used to enhance health research processes and administrative tasks in health institutions; however, their use for these purposes is limited due to different factors such as confidentiality, heterogeneity of information and incompleteness of medical data. As a consequence, most of decision-making in health sector does not take full advantage of the vast source of information from EHR because of the difficulty of obtaining the adequate information at the right moment. This section aims to analyze how can be improved the decision making process specifically for clinical research and health administration using as a source of information EHR.

The problems around decision-making were studied in a general hospital in Bogotá, Colombia. This hospital has a main information system called SAHI that includes modules to manage the Electronic Health Record, Contracts, Human Resources, Client Service, Budgeting, Purchases and Supply, etc. This system allows physicians to obtain the EHR of each patient during the medical attention; however, their use for research purposes has been limited due to the fact that important information is stored in narrative texts, intended for human beings that are difficult to search and analyze automatically. One of the common requirements of medical research is to find the medical records of patients that have been diagnosed with a specific disease. This task that should be easily done using classical queries (e.g. using SQL) is very time-consuming. This is because diagnosis is frequently hidden in narrative texts (e.g. medical notes, progress notes), hindering the possibility of automatically detecting relevant records and requiring the participation of an expert in the analysis. Similarly, the administration of the Hospital frequently requires analyzing costs and efficiency of medical treatments. Even if some of this information is well structured, as medications and laboratory orders, the complete sequence of events related to a patient is hidden in narrative texts. In summary, the main requirement of this hospital decision support system is to recognize which medical records are useful for clinical and administrative research, taking into account all the information in the EHR, including the one that is in narrative texts. Particularly, the decision support system must support biomedical research and quality analysis and service delivery. In the following, both of them are going to be explained.

3.1 Biomedical Research

Biomedical research is an approach used for solving medical problems based on proving theories or hypothesis through observations and experimentations. The clinical information contained in EHR, can be useful in different ways in this type of research depending on the selected research design [27]. Below are some of the research designs most frequently used, and the way in which the data contained in the EHR could be useful.

Descriptive Studies. The objective in these studies is to describe the variables in a group of subjects during a short period of time, not including the control group. The variables to describe typically include age, gender, occupation and lifestyle. These studies are also useful to describe variables related to the time factor, as the frequency of the disease during different periods of time. For health administrators these studies let describing the general characteristics of the disease distribution, allowing a better prioritization of resources. For researchers, these studies are the first step to study possible risk factors, which are very useful for formulating hypotheses that can be tested subsequently through an analytical design. Descriptive studies include case reports, case series, cross-sectional studies and correlational or population studies. Normally these designs are used for describing uncommon cases, seeking to generate hypotheses that will encourage further studies. However, conducting descriptive studies typically requires identifying patients with the condition or disease under study. Automated record obtainment usually relies on coded attributes (e.g. using CIE 10) which is not only complex to fill in at record time, but also limited in actual research usage [28, 29].

Analytical Studies. Analytical studies allow researchers to evaluate the effects of an intervention or risk factor. Essentially, these studies compare the frequency of outcomes in the intervention group with the control group, thus determining whether the observed outcome depends on the evaluation factor. The design of these studies allows to define and to evaluate a hypothesis. Analytical studies include case–control studies, cohort studies and clinical experiments. These types of studies are increasingly being used, for instance to check adherence to treatment protocols (e.g. [30]). For both kinds of research designs it is required to recognize patients who have specific characteristics of interest. In some cases, as in study cases, it must be identified the presence of an uncommon disease or a population with specific characteristics. In the case of a retrospective cohort study, it should be recognized a group of patients who have had exposure to a given risk factor at a given time. A decision support system could facilitate this task by allowing the identification of patients who meet the inclusion and exclusion criteria defined by the researcher, who will decide the characteristics such as gender, age group, diagnosis or exposure to a risk factor.

3.2 Quality and Service Delivery Analysis

The main goals of EHR systems are to enhance coordination between different health care participants and to promote the use of clinical guidelines, which would improve, therefore, the overall quality of care. Particularly, a systematic use of EHR contents could improve quality and cost-effectiveness analysis over treatments and medical decisions by using the evidences registered on patients with specific characteristics. Having accurate information on the characteristics of the patients, both in their epidemiology and severity aspects, as well as the most frequently activities and procedures executed by medical doctors, medical institutions and comparing them with final outcomes will allow to improve not only medical-care processes, but at the same time the efficiency on the use of medical related resources.

One specific requirement to improve decision making from the service delivery point of view involves assessing adherence to international recommendations or guidelines of patient treatments. Evaluating the effectiveness of a drug or recommendation could be improved through the measurement of quality using an analytic decision support system. The results of this kind of analysis can act as an input to modify hospital internal policies as well as justifying changes over government policies around health services. Currently, the processes for assessing adherence to international recommendations and in general to evaluate medical decisions include long term activities of analysis of EHR. Figure 2 illustrates a classic EHR analysis process from the point of view of data requirements. As it can be seen an important and currently time-consuming task is the exploration and identification of relevant EHR. IT staff members are the owners of the EHR and they are frequently a bottleneck during research projects. In addition, the effort to

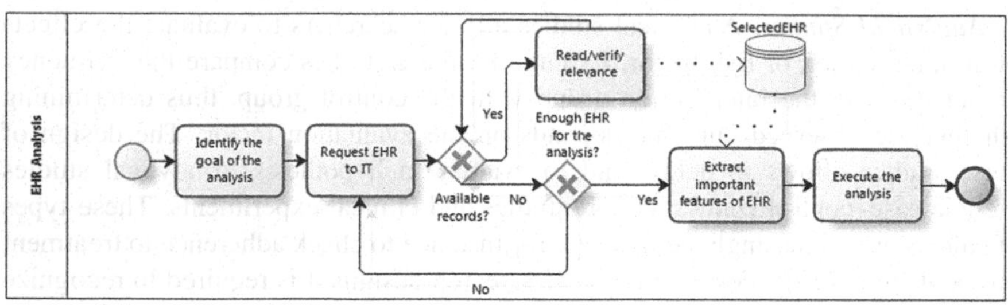

Fig. 2 HER analysis process

validate the relevance of the provided EHR implies time between 5 and 40 min for reading each one of the attentions of patients. Providing a facility to obtain these EHR automatically will improve process time metrics. In addition, providing visual analysis of the relevant EHR enriches the quality of the decision process due to the improvement of user empowerment.

4 DISEArch: A System for Electronic Medical Record Analysis

This section describes the decision support system created in order to support the requirements of clinical research and health administration, mentioned in the previous section, using as a source of information EHR. The main principle of this system called DISearch is to combine the analysis of structured and unstructured information contained in EHR to enhance decision-making. Considering the main users of DISEArch are medical doctors, during the development of this system the definition of functionalities and the user interface were made using a participatory design approach. In this approach users cooperate with designers and developers during the different phases of the project. The design followed an evolutionary cycling where the system was discussed mainly from a clinical and practical point of view rather than a technical perspective. In what follows, this presents the general architecture of the system and explains in detail its main components.

4.1 DISEArch Architecture

The architecture of DISEArch is illustrated in Fig. 3. The principle of the system is to provide different capabilities of visualization and analysis to enhance biomedical research as well as quality and services delivery analysis. The components are divided in three layers. The components of the Data Layer are in charge of store all the information used or generated by the system. The *Knowledge base* component

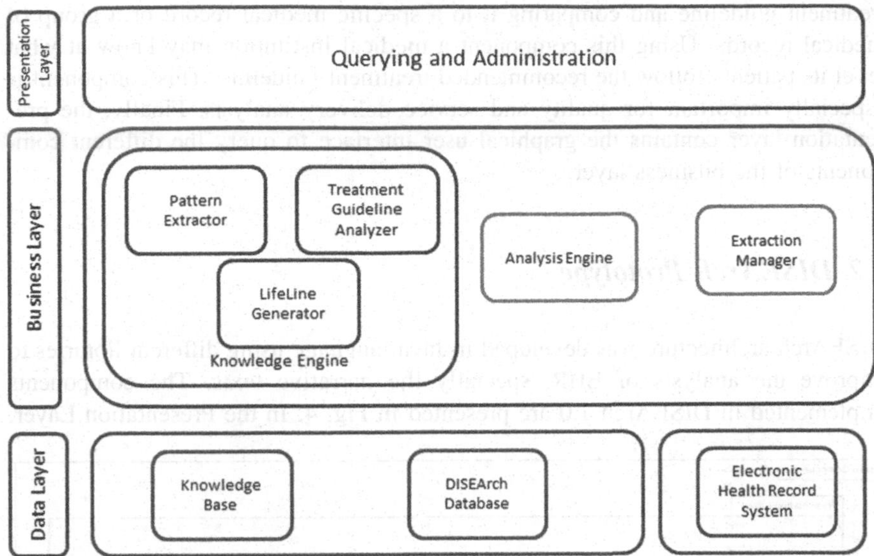

Fig. 3 DISEArch architecture

manages the relevant taxonomies and definitions required to analyzed EHR. For instance, it stores the taxonomies of diseases and medications that are used to categorize a record. The *DISEarch database* stores all the temporal results and the analysis results obtained by the components of the business level. The *Electronic Health System* component represents the database of the Hospital Information System that contains the EHR. Even if this last component is external, DISEArch contains all the information required to extract the EHR from it.

The Business Layer components contain the main capabilities of the decision system. The *Extraction Manager* is responsible for extracting the set of medical records according to the pre-selection defined by the users (e.g. gender, age, date). This is the only component that interacts directly with the *Electronic Health System* component and contains the logic for wrapping the schema of the original source of health records. In addition, when it is required, this component launches an anonymization service that assures the de-identification of identifiers and seudo-identifiers from EHR. The *Analysis Engine* is in charge of analyzing the structured and non-structured elements of EHR to identify and prioritize which of them are useful for a specific study (e.g. descriptive study). The *Knowledge Engine* goal is to extract knowledge from different points of view. First, the *LifeLine Generator* presents a 360-degree view of a patient visualizing all the events where he/she has been involved. The event extraction includes the identification of events that were written in narrative text as well as those that are well structured. The second component is the *Pattern Extractor*; this component applies data mining techniques over a set of EHR to obtain natural clusters, association rules and sequence patterns. The *Treatment Guideline Analyzer* allows structuring a disease treatment guideline and comparing it to a specific medical record or a group of medical records. Using this component a medical institution may know at what level its patients follow the recommended treatment guideline. This component is especially important for quality and service delivery analysis. Finally, the presentation layer contains the graphical user interface to query the different components of the business layer.

4.2 DISEArch Prototype

DISEArch architecture was developed in Java language using different libraries to improve the analysis of EHR, specially the narrative texts. The components implemented in DISEArch 1.0 are presented in Fig. 4. In the Presentation Layer,

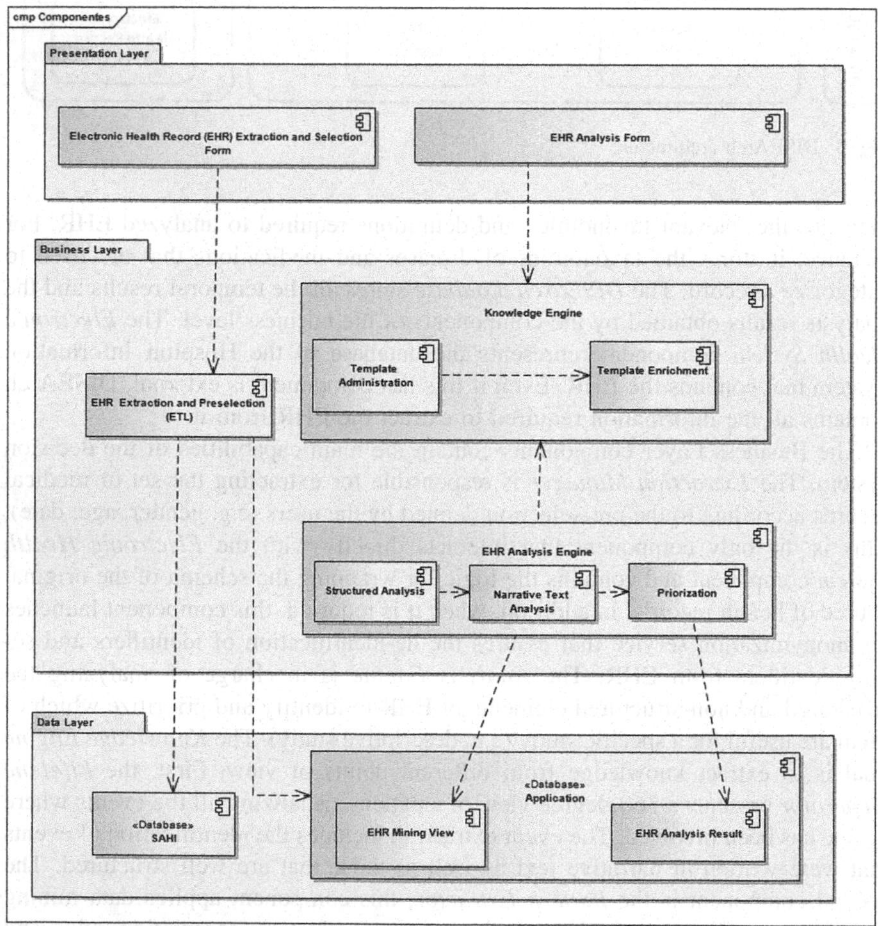

Fig. 4 DISEArch prototype

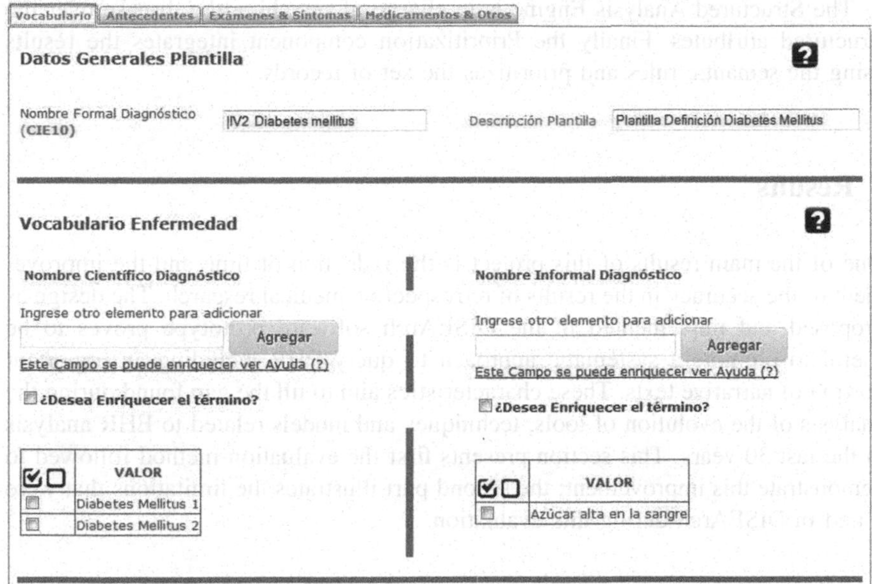

Fig. 5 DISEArch presentation

the researcher can describe the characteristics of the EHR that he/she is interested on using the EHR analysis form. As presented in Fig. 5, this form allows specifying the characteristics of the EHR required for the analysis like the vocabulary of the disease, the related lab exams, etc. The Knowledge Engine handles the knowledge base that allows the enrichment of the description of the disease defined in the EHR analysis form. This enrichment increases the definitions made by the researcher adding synonyms or related words to the initial description.

The Extraction and Pre-selection component is in charge of the extraction and initial preprocessing of medical records from the EHR system. This component is parameterized according to the characteristics of the system and extracts the records according to the definition of initial parameters, such as date of admission, gender or age of patients. This component was built using the analysis services of SQL Server 2008.

The EHR Analysis Engine is the core of the analysis. It uses natural language processing to analyze EHR. It implements Stemming using Porter Stemmer algorithm, simple string tokenization, sentence splitting, POS tagging using Probabilistic Part-of-Speech Tagging Using Decision Trees [31] for annotating text with part-of-speech and lemma information and finally gazetteer lookup using regular expressions. This component has a coordinator that calls each of the search engines. The Narrative Text Analysis is in charge of the analysis of natural language and was developed using the GATE API [32]. This API enables the inclusion of all the language processing functionality within DISEArch. In addition, we use Tree tagger [31], a Pearl implementation which provides tokenization and Part of the Speech tagger.

The Structured Analysis Engine is in charge of searching the disease over the structured attributes. Finally the Prioritization component integrates the results using the semantic rules and prioritizes the set of records.

5 Results

One of the main results of this project is the reduction of time and the improvement of the accuracy in the results of retrospective medical research. The design as proposed and implemented in the DISEArch software prototype proves to be useful to promote a systematic approach to query EHR including an important portion of narrative texts. These characteristics aim to fill the gap found during the analysis of the evolution of tools, techniques and models related to EHR analysis in the last 30 years. This section presents first the evaluation method followed to demonstrate this improvement; the second part illustrates the limitations that were found in DISEArch during this evaluation.

5.1 Evaluation Results

The aim of this project is to allow the hospital administrators and medical researchers to recognize the utility of the information contained in EHRs to improve quality of service and research processes. To evaluate this goal we developed a process to measure the utility of the system, the quality of the results (in terms of the accuracy of obtained records) and the efficacy of the system to change the related processes. The team that carried out the evaluation combined medical doctors, students of medical specializations and the university hospital IT staff. For evaluation purposes, three diseases were selected as case studies: diabetes mellitus, heart failure and pulmonary hypertension. These diseases were selected taking into consideration their common characteristic of being chronic diseases. This alone makes the use and evaluation of the system easier to analyze, because it is more likely that such patients have continued visits to the hospital, thus making their EHRs a richer source of data. In addition, those diseases also exhibit differences related to their ease of detection, their prevalence and incidence proportions, which helps in triangulating the evaluation results. The utility and quality of results were measured using the EHR from the SAHI system. During the first phase a group of medical doctors analyzed manually 400 EHRs selected randomly from patients in the department of Internal Medicine. From this analysis we obtained three sets of EHR, each one with the patients with the selected diseases. In essence, this is used as the control dataset, obtained with the current manual method for obtaining EHRs related to a disease.

The second phase consisted in the use of DISEARch by a group of medical doctors specialized in each one of the diseases. They described the diseases using

Table 2 Utility evaluation results

	Manual analysis		
		Positive	Negative
DISEARch	Positive	70	1
	Negative	2	316

DISEArch and ask for the relevant EHRs. The utility was measured evaluating the consistency of the manual selection results with respect to the selected EHRs by DISEArch. The results of this evaluation are presented in Table 2. It shows the number of EHR that were identified manually as positive or negative in comparison with the detection of DISEArch.

As it can be seen, the number of Negative–Positives and Positive–Negatives are considerably lower with respect to the number of Positive–Positive and Positive–Negative. These results prove the utility of DISEARch to identify the EHR required for the analysis. However, even if lower, the not-matching results were analyzed to identify where and why automatic detection was not possible. Medical doctors analyzed once again all the events associated with the "problematic" EHRs. In the case of the Positive–Negative EHR, DISEArch detected a patient relevant for the analysis of Heart Failure; on the contrary, medical doctors determined that the patient did not have the disease. The new analysis identified that it was a patient with an early stage of Heart Failure and because of that the manual analysis determined it was not relevant; however, the evolution of the patient would probably end in a Heart Failure diagnosis. According to the doctors, neither the medical doctor that did the early analysis, nor DISEarch were wrong: the difference was the level of Heart Failure evolution that each one of them aimed to used to identify EHR. This is worth noting because neither the manual process nor DISEArch are free from error. In fact, as discussed in an evaluation meeting, some medical doctors view the situation as if each individual actually carries all diseases potentially in him or her, and if we lived long enough we would all eventually develop the diseases. As such, selecting patients will always leave out some that could have been included and it is a human decision whether to include those with borderline conditions. The two Negative-Positive EHR detected correspond to patients with a large number of service records (visits, tests, hospitalizations, etc.). Some of these included early medical statements discarding the disease but more recent ones confirmed its presence. These cases represent a great challenge to DISEArch because the analysis involves a new variable of analysis that initially was not taken into account; the next section develops this issue further.

The efficacy of the system to renew the process of analysis was measured comparing the previous (manual) process with the new (DISEarch-supported) one, and the effects it has in time invested and user empowerment. Figure 6 illustrates the new process. In this new process the final user is not dependent on IT staff, avoiding the bottleneck that this implied; In addition, Table 3 illustrates improvements on the time required for the analysis of EHR.

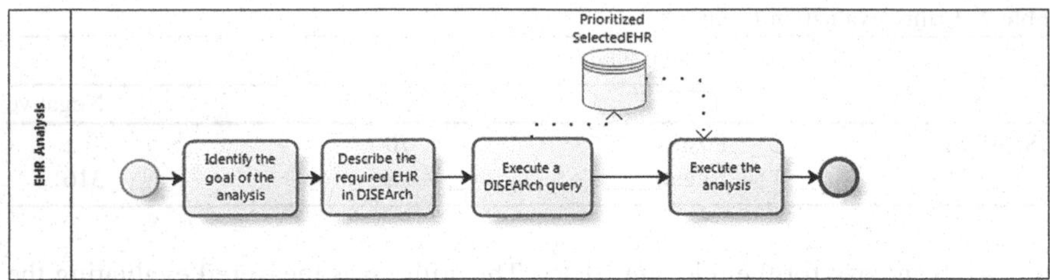

Fig. 6 EHR analysis process with DISEArch

Table 3 DISEarch time performance

	Manual process	DISEArch
Number of HER	1350	400
Time required	10 min	166 h

5.2 DISEArch Challenges

During the evaluation of DISEArch we detected two short-term challenges: the identification of contrary narrative texts from EHR and the improvement of performance of analytical components. Long-term challenges are discussed in Sect. 4.

The first challenge involves what we have called "recognition of pattern revocation", it means that DISEArch must identify patterns discarding a disease in a patient from narrative and structured attributes and at the same time it must identify patterns that invalidate the previous one over time. Consider a patient with an EHR with 30 services or visits, the first 15 may involve the search for a disease X that finally is discarded in service number 15; from 16 to 21, the disease X is not relevant for medical doctors and because of that these service records do not have any information related to it; finally from 22 to 29, disease symptoms reappear and the disease was confirmed at service record 30. This case requires the identification of both facts, the one that discarded the disease and the one that confirmed the disease, and the application of a function to identify which one has the higher weight to confirm or deny the inclusion of the EHR in the final results.

The second challenge is more technical and implies the improvement of performance of the services that analyzes the EHR pieces of information. To accomplish this improvement, it is necessary make a balance between flexibility and efficiency of DISEArch.

5.3 Implications for IT-Based Service Systems

In order to assure that the new system would deliver the expected results, the project followed the guidelines of the MOF framework [33]. The *Plan*, *Deliver*, *Operate* and *Manage* phases of the IT service lifecycle allowed us to bring together information and people; however, in some of these phases we had to take important decisions that worth mentioning.

During the *Plan* phase, the formation of the project team was extremely time consuming because of the misalignment between the IT area and the legal and human resources department. Even if these problems were solved, for new projects it will be essential to do a pre-analysis of the financial structure of the project to prevent time consumption in administrative tasks. Considering this project followed a participatory design approach, during the *Manage* phase we needed to take into account an open strategy for change management. In this strategy we defined formal mechanisms to register required changes on the system, but an open participatory mechanism to evaluate the consequences on time and effort to deliver these changes. IT analysts worked together with final users to prioritize changes according to the effects it will have on the improvement of the business value. In such a way, we avoided frequent iterations that will impact project time. Since the initial problem on this project was ill-structured, the Envision function of the *Deliver* phase did not include a clearly documented vision and scope of the project. Nevertheless, the fact that the project followed a participatory design approach allowed to reduce the risks of this shallow original definition.

5.4 Discussion

In fact, a long term goal of this research is to gradually improve the quality of patient treatments based on formal evidence gathered through the specific analysis of each disease, especially chronic diseases.

The process used to design and create DISEArch followed an integration of rigor and relevance reinforced through user experience and involvement during the design that enabled the appropriation and evolution of the information system created. This approach demonstrates the value of participatory design to enrich the requirements phase during all the process to assure the quality of the final design and the fulfillment of user expectations. However, the drawback of the approach is the high time consumed to undertake the process, particulary during the design of the user interface. This may not be a disadvantage but the evidence that time typically used for this activity was considerably lower than required.

6 Conclusions

This chapter presents a health decision-support system called DISEArch that allows the identification and analysis of relevant EHR for decision-making. It uses structured and non-structured data, and provides analytical as well as visualization facilities over individual or sets of EHR. DISEArch proves to be useful to empower researchers during analysis processes and to reduce considerably the time required to obtain relevant EHR for a study. This work opens important research issues. Among them, the extension of DISEARch design to include temporal analysis over events of EHR with the same diagnosis. This kind of analysis involves the recognition of events from narrative texts and the application of time series analysis to identify temporal patterns. Such event-based patterns should contribute to matching actual treatment against proposed protocols and guidelines, open up traceability for a clinical and administrative use, and provide alternative views on health records that can enhance daily medical practice. In addition, the use of text mining techniques for creating abstract summaries of EHR in DISEArch is promising. The combination of such techniques with the analysis of semantic distance between EHR should also be further developed.

Future research also involves the use of our proposal in other medical institutions leading probably to other analysis requirements and technical issues related to other hospital information systems. Probably the most important research perspective we consider, is the further use of participatory design for the enrichment of health decision-making system. It is complex but necessary to strike a balance and generate a fruitful discussion from very different sets of expertise. As with any information systems project, a conversation needs to be put in place to realize the full potential that IT-based systems offer for people, in this case within the medical domain. It is a mutual learning experience that requires constant translations, frequent prototype discussions, grounding of new IT-based support in current practices and clear identification of existing problems and future opportunities that are opened up in order to enrich the momentum of the project, enlarge the community of early adopters and guaranteeing the continued financial, scientific and administrative support for the project from management stakeholders. Our experience is very positive and we intend to further pursue this approach and extract lessons learned for similar projects.

Acknowledgments This work is part of the project entitled "Identificación semiautomática de pacientes con enfermedades crónicas a partir de la exploración retrospectiva de las historias clínicas electrónicas registradas en el sistema SAHI del Hospital San Ignacio" funded by Hospital Universitario San Ignacio and Pontificia Universidad Javeriana.

References

1. Gresh, D.L., Rabenhorst, D.A., Shabo, A., Slavin, S.: PRIMA: A case study of using information visualization techniques for patient record analysis. In: Proceedings of the IEEE Visualization, pp. 509–512 (2002)
2. Lisboa, P.J.G.: A review of evidence of health benefit from artificial neural networks in medical intervention. Neural Netw. **15**, 11–39 (2002)
3. McCauley, N., Ala, M.: The use of expert systems in the healthcare industry. Inform. Manage. **22**, 227–235 (1992)
4. Kalim, K., Carson, E., Cramp, D.: The role of soft systems methodology in healthcare policy provision and decision support. In: Proceedings of the IEEE International Conference on Systems, Man and Cybernetics, pp. 5025–5030 (2004)

5. Kalim, K., Carson, E.R., Cramp, D.: An illustration of whole systems thinking. Health Serv. Manage. Res. **19**, 174–185 (2006)
6. Hesse, B.W., Shneiderman, B.: eHealth research from the user's perspective. Am. J. Prev. Med. **32**, S97–S103 (2007)
7. Carroll, J.M., Rosson, M.B.: Participatory design in community informatics. Des. Stud. **28**, 243–261 (2007)
8. Mao, J.-Y., Vredenburg, K., Smith, P.W., Carey, T.: The state of user-centered design practice. Commun. ACM **48**, 105–109 (2005)
9. Bellazzi, R., Zupan, B.: Predictive data mining in clinical medicine: current issues and guidelines. Int. J. Med. Inform. **77**, 81–97 (2008)
10. Windle, P.E.: Data mining: an excellent research tool. J. Perianesthesia Nurs.: Official J. Am. Soc. PeriAnesthesia Nurs./Am. Soc. Perianesthesia Nurs. **19**, 355–356 (2004)
11. Hevner, A.R., March, S.T., Park, J., Ram, S.: Design science in information systems research. MIS Q. **28**, 75–105 (2004)
12. Hevner, A.R.: A three cycle view of design science research. Scand. J. Inform. Syst. **19**, 39–64 (2007)
13. Shearer, C.: The CRISP-DM Model: The new blueprint for data mining. J. Data Warehouse. **5**, 13–22 (2000)
14. Gaspar, L., Scott, C., Rotman, M., Asbell, S., Phillips, T., Wasserman, T., McKenna, W.G., Byhardt, R.: Recursive partitioning analysis (RPA) of prognostic factors in three radiation therapy oncology group (RTOG) brain metastases trials. Int. J. Radiat. Oncol. Biol. Phys. **37**, 745–751 (1997)
15. Highet, V.S., Forrest, A., Ballow, C.H., Schentag, J.J.: Antibiotic dosing issues in lower respiratory tract infection: population-derived area under inhibitory curve is predictive of efficacy. J. Antimicrob. Chemother. **43**, 55–63 (1999)
16. Burbidge, R., Trotter, M., Buxton, B., Holden, S.: Drug design by machine learning: Support vector machines for pharmaceutical data analysis. Comput. Chem. **26**, 5–14 (2001)
17. Kroes, R., Renwick, A.G., Cheeseman, M., Kleiner, J., Mangelsdorf, I., Piersma, A., Schilter, B., Schlatter, J., Van Schothorst, F., Vos, J.G., Würtzen, G.: Structure-based thresholds of toxicological concern (TTC): Guidance for application to substances present at low levels in the diet. Food Chem. Toxicol. **42**, 65–83 (2004)
18. Suppes, T., Dennehy, E.B., Hirschfeld, R.M.A., Altshuler, L.L., Bowden, C.L., Calabrese, J.R., Crismon, M.L., Ketter, T.A., Sachs, G.S., Swann, A.C.: The Texas Implementation of Medication Algorithms: Update to the algorithms for treatment of bipolar I disorder. J. Clin. Psychiatry **66**, 870–886 (2005)
19. Hill, R.P., Lubarsky, D.A., Phillips-Bute, B., Fortney, J.T., Creed, M.R., Glass, P.S.A., Gan, T.J.: Cost-effectiveness of prophylactic antiemetic therapy with ondansetron, droperidol, or placebo. Anesthesiology **92**, 958–967 (2000)
20. Adams, P.C., Gregor, J.C., Kertesz, A.E., Valberg, L.S.: Screening blood donors for hereditary hemochromatosis: Decision analysis model based on a 30-year database. Gastroenterology **109**, 177–188 (1995)
21. Hess, K.R., Abbruzzese, M.C., Lenzi, R., Raber, M.N., Abbruzzese, J.L.: Classification and regression tree analysis of 1000 consecutive patients with unknown primary carcinoma. Clin. Cancer Res. **5**, 3403–3410 (1999)
22. Wang, T.D., Wongsuphasawat, K., Plaisant, C., Shneiderman, B.: Extracting insights from electronic health records: case studies, a visual analytics process model, and design recommendations. J. Med. Syst. **35**, 1135–1152 (2011)
23. Thiessard, F., Mougin, F., Diallo, G., Jouhet, V., Cossin, S., Garcelon, N., Campillo, B., Jouini, W., Grosjean, J., Massari, P., Griffon, N., Dupuch, M., Tayalati, F., Dugas, E., Balvet, A., Grabar, N., Pereira, S., Frandji, B., Darmoni, S., Cuggia, M.: RAVEL: retrieval and visualization in electronic health records. Stud. Health Technol. Inform. **180**, 194–198 (2012)
24. Hsu, W., Taira, R.K., El-Saden, S., Kangarloo, H., Bui, A.A.T.: Context-based electronic health record: toward patient specific healthcare. IEEE Trans. Inform. Technol. Biomed.: Publi. IEEE Eng. Med. Biol. Soc. **16**, 228–234 (2012)

25. Muller, H., Maurer, H., Reihs, R., Sauer, S., Zatloukal, K.: Adaptive visual symbols for personal health records. In: Proceedings of the 2011 15th International Conference on Information Visualisation (IV), pp. 220–225 (2011)
26. Bui, A.A.T., Aberle, D.R., Kangarloo, H.: TimeLine: visualizing integrated patient records, IEEE Trans. Inform. Technol. Biomed.: Publi. IEEE Eng. Med. Biol. Soc. **11**, 462–473 (2007)
27. Arguedas-Arguedas, O.: Tipos de diseño en estudios de investigación biomédica. Acta Médica Costarricense **52**, 16–18 (2010)
28. Calandre-Hoenigsfeld, L., Bermejo-Pareja, F.: Difficult-to-classify symptoms and syndromes in a series of 5398 neurological outpatients diagnosed according to ICD-10 criteria. Síntomas y síndromes de difícil clasifcación en una serie ambulatoria de 5.398 pacientes neurológicos diagnosticados según la CIE-10 **53**, 513–523 (2011)
29. Tanno, L.K., Ganem, F., Demoly, P., Toscano, C.M., Bierrenbach, A.L.: Undernotification of anaphylaxis deaths in Brazil due to difficult coding under the ICD-10. Allergy: Eur. J. Allergy Clin. Immunol. **67**, 783–789 (2012)
30. Komajda, M., Lapuerta, P., Hermans, N., Gonzalez-Juanatey, J.R., Van Veldhuisen, D.J., Erdmann, E., Tavazzi, L., Poole-Wilson, P., Le Pen, C.: Adherence to guidelines is a predictor of outcome in chronic heart failure: the MAHLER survey. Eur. Heart J. **26**, 1653–1659 (2005)
31. Schmid, H.: Probabilistic Part-of-Speech Tagging Using Decision Trees. In: Proceedings of the International Conference on New Methods in Language Processing, Manchester, UK, (1994)
32. Cunningham, H., Maynard, D., Bontcheva, K., Tablan, V., Aswani, N., Roberts, I., Gorrell, G., Funk, A., Roberts, A., Damljanovic, D., Heitz, T., Greenwood, M., Saggion, H., Petrak, J., Li, Y., Peters, W., Others: Text Processing with GATE (Version 6), University of Sheffield Department of Computer Science Sheffield (2011)
33. Leenards, P., Pultorak, D., Henry, C.: MOF: V4.0, (Microsoft Operations Framework 4.0): Version 4.0: A Pocket Guide, Van Haren Publishing (2008)

Chapter 10
Architecture for Business Intelligence Design on the IT Service Management Scope

C. P. Marin Ortega, C. P. Pérez Lorences and -Ing. Habil J. Marx-Gómez

Abstract In the present research we propose new business intelligence architecture to support the IT balanced scorecard cascade based on the integration of business and technological domains for the IT service management. This paper presents some preliminary results on the state of the art analysis on the topics: IT BSC, business intelligence, and aggregation methods based on the fuzzy logic operators to build management indicators. The main contributions are: new architecture for business intelligence design, an aggregation method to design new indicators based on the statistic and compensatory fuzzy logic approach taking into account as sources the indicators defined in the COBIT and ITIL frameworks. As the first result we define a new IT BSC for the Cuban enterprise.

Keywords Business intelligence · IT governance · IT services management balanced scorecard · IT balanced scorecard · COBIT · ITIL · Aggregation methods

1 Introduction

The increasing use of information management technologies within organization has resulted in Information Technology (IT) usage-dependant organizations seeking to have increasingly efficient and innovative technological services and

---

C. P. Marin Ortega (✉) · C. P. Pérez Lorences
Department of Industrial Engineering, Central University of Las Villas, Santa Clara, Cuba
e-mail: pablomo@uclv.edu.cu

C. P. Pérez Lorences
e-mail: patriciapl@uclv.edu.cu

-Ing. H. J. Marx-Gómez
Department of Computing Science, Business Information Systems I/VLBA, Carl von Ossietzky University Oldenburg, Oldenburg, Germany
e-mail: jorge.marx.gomez@wi-ol.de

M. Mora et al. (eds.), *Engineering and Management of IT-based Service Systems*, Intelligent Systems Reference Library 55, DOI: 10.1007/978-3-642-39928-2_10,

solutions. Organizations recognize that IT services are strategic assets to support information and service management. However, the reality is often that these services are overlooked or not addressed at all, with the strategic importance they entail [1]. Organizations should be aware of the close relationship and convergence between business and IT. However, integration of business needs and technology is still a challenging issue. As a response to this problem, IT Service Management (ITSM) enables the integration of business with IT in terms of services that can be managed as another business unit. IT services are recognized as crucial, strategic, organizational assets that must be managed for business success [2].

A management system based on the strategy is a tool that helps the manager discovering what is really important in order to achieve the company's goals. The balanced scorecard (BSC) initially developed by Kaplan and Norton, is a performance management system that enables businesses to derive strategies based on measurement and follow-up. In recent years, the BSC has been applied to information technology. The IT BSC is becoming a popular tool to enable the integration of business with IT.

In the ITSM context have been developed frameworks such as COBIT [3] and ITIL [4] that could be used for the selection of metrics to build up the IT BSC in an organization. However, there is a lot of information, COBIT for example has 34 processes and it provides more than ten Key Performance Indicators (KPI) and Key Goal Indicators (KGI) for each one.

The business intelligence (BI) tools alone facilitate "how" it is possible to achieve a solution from the informatics point of view, but they do not assure, "what" is the information that is really needed. Often occurs, that parts of the whole information needed to make decisions is missing. The reason for that is either lack of knowledge from the person requesting it, or because the information is not supported in the organization's database. To improve it the BI design must take into account the information related to the KPI or KGI defined in the COBIT and ITIL frameworks. Additionally the absence in these frameworks of indicators capable of monitoring the strategy in a more integral way and most of the information needed for its design is essentially based on subjective and imprecise concepts expressed primarily by "experts" in a natural language are nowadays gaps in the science.

Compensatory Fuzzy Logic (CFL) is an area that can fill these gaps largely because it uses language as communication, it creates an explicit model of preferential knowledge and then uses the inference capability of the logical platform to propose decisions that better reflect the decision policy of individuals [5].

To merge CFL and the aggregation methods approach to create new indicators for BSC and IT BSC helps us to summarize a lot of information in a few indicators, and it will be a good starting point to design the fact table in a data warehouse.

This paper focuses on the state of the art analysis on the BSC and IT BSC approach, BI components, aggregation methods and compensatory fuzzy logic approach. Furthermore we propose new architecture for business intelligence design based on the mapping among the business strategy, COBIT and ITIL frameworks in order to define new indicators for the BSC and IT BSC based on the indicators defined in the aforementioned frameworks.

This paper is organized as follows: Sect. 2 provides related work. In Sect. 3, we present important outlines about the new architecture. Section 4 we show some partial results. Finally, conclusions and future work are discussed in Sect. 5.

2 Related Word

In the early 1990s, Kaplan and Norton introduced the BSC at enterprise level. Their fundamental premise is that the evaluation of a firm should not be restricted to a traditional financial evaluation but should be supplemented with objectives and measures concerning customer satisfaction, internal processes and the ability to innovate.

The BSC can be applied to the IT function, its processes and projects. Gold [6] and Willcocks [7] have conceptually described and has been further developed by Van Grembergen and Van Bruggen [8], Van Grembergen and Timmerman [9] and Van Grembergen [10].

To achieve that, the focus of the four perspectives of the business BSC needs to be translated as shown in Table 1. The User Orientation perspective represents the user (internal or external) evaluation of IT. The Operational Excellence perspective represents the IT processes employed to develop and deliver the applications. The Future Orientation perspective represents the human and technology resources needed by IT to deliver its services over time. The Business Contribution perspective captures the business value created from the IT investments.

Each of these perspectives has to be translated into corresponding goals and metrics that assess the current situation. These assessments need to be repeated periodically and aligned with pre-established goals and benchmarks [11]. The IT BSC links with the business, mainly through the business contribution perspective. The relationship between IT and business can be more explicitly expressed through a cascade of scorecards [10, 12]. In Fig. 1 is showed the relationship between IT scorecards and the business scorecard. The IT Development BSC and the IT Operational BSC both are enablers of the IT Strategic BSC that in turn is the enabler of the Business BSC. This cascade of scorecards becomes a linked set of measures that will be instrumental in aligning IT and business strategy and that will help to determine how business value is created through IT.

The case under review in [13] illustrated one of the most crucial issues in building and implementing an IT strategic BSC: its required linkage with the business objectives. To create this link a cascade of BSC has been established with the unit scorecards at a lower level for the operational and development services. The measures of these unit scorecards are rolled-up or aggregated in the IT strategic scorecard that ultimately realizes the link with the business objectives through its corporate contribution perspective. The precise articulation of the cause-and-effect relationships through the identification of outcome measures and

Table 1 Generic IT BSC

User orientation	Business contribution
How do users view the IT department?	How does management view the IT department?
Mission	Mission
To be the preferred supplier of information systems.	To obtain a reasonable business contribution from IT investments.
Objectives	Objectives
• Preferred IT supplier	• Control of IT expenses
• Partnership with users	• Business value of the IT function
• User satisfaction	• Business value of new IT projects
Operational excellence	Future orientation
How effective and efficient are the IT processes?	How well is IT positioned to answer future challenges?
Mission	Mission
To deliver effective and efficient IT applications and services.	To develop opportunities to answer future challenges.
Objectives	Objectives
• Efficient computer operations	• Training and education of IT staff
• Efficient help desk function	• Expertise of IT staff
• Efficient software developments	• Research into emerging information technologies

Fig. 1 BSC cascade

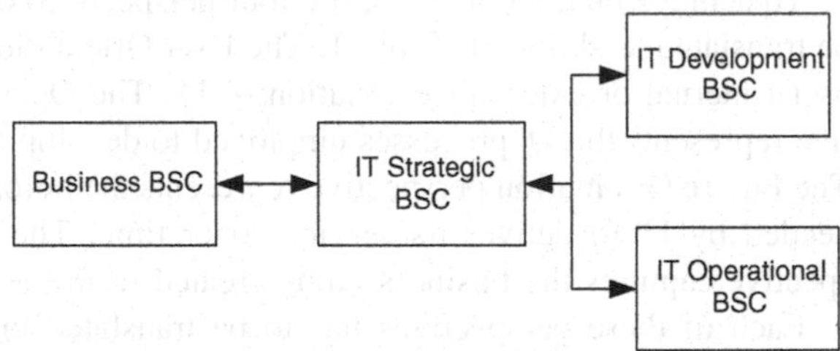

their corresponding performance drivers seemed to be a critical success factor. These relationships are implicit in the current IT strategic BSC but are to be defined more explicitly.

The BI tools are the best choice to support the BSC cascade and show it to the business people, because it is used to provide historical, current, and predictive views of business operations. Common functions of BI technologies are reporting, on-line analytical process (OLAP), analytics, data mining, business performance management, benchmarks, text mining, and predictive analytics [14].

Generally, a BI system should have the following basic features [15]:

- Data Management: including data extraction, data cleaning, data integration, as well as efficient storage and maintenance of large amounts of data.
- Data Analysis: including information queries, report generation, and data visualization functions.
- Knowledge Discovery: extracting useful information (knowledge) from the rapidly growing volumes of digital data in databases.

Nowadays the fuzzy operators are widely used in the decision making field [16–20] because some of their essential properties including the capability to properly model the "vagueness" of natural language and uncertainty. Based on this strength the business analyst can builds a semantic model and then they can use the inference capability of the logical platform to facilitate assessment on any situation, decision making or knowledge discovery. In [21] appears an state-of-the-art overview on aggregation theory based on fuzzy logic, also in [22–24] is showed the aggregation methods approach. We propose in this research the use the aforementioned methods to define the indicators in the BSC cascade based on best practice frameworks for IT service management such as COBIT and ITIL.

3 Architecture for Business Intelligence Design on the IT Service Management Scope

In accordance with the previously expressed, the proposed architecture is comprised by the Enterprise Scope, Management Control Tools, BI components and relationships among them as shown in Fig. 2

The Enterprise Scope is compound by the business scope, the COBIT and ITIL frameworks. Business Strategy including mission, vision, the strategic goals and management plan for the business.

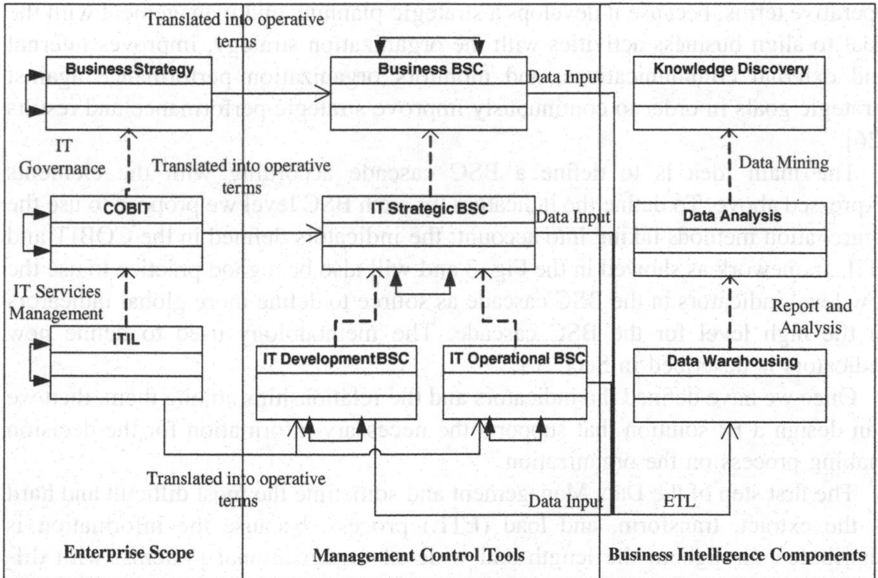

Fig. 2 Business intelligence architecture on the IT service management scope

COBIT is a globally accepted set of tools organized into a framework that ensures IT is working as effectively as possible and maximizes the benefits of technology investment. COBIT acts as an integrator of more detailed international IT standards and guidance. Based on industry standards and best practices, it is a proactive, uniquely comprehensive approach to ensure that IT is meeting the needs of an enterprise and enabling the achievement of strategic business objectives. It easily integrates with and builds on other business and IT frameworks and standards (such as COSO, ITIL and ISO 27001), while improving their impact.

ITIL is a series of books and is referred to as the only consistent and comprehensive best practice for IT service management to deliver high-quality IT services. Although produced and published by a single governmental body, ITIL is not a standard. ITIL is usually implemented subject to one or more of the following business cases to: define service processes within the IT organization, define and improve the quality of services, focus on the IT customer or implement a central help desk function. As mentioned previously, ITIL is a collection of books related to IT service management, however does not describe the what; it is focused on how and who.

According with [25] it is possible mapping COBIT and ITIL frameworks. The mapping is performed in two levels. A high-level mapping compares the objectives stated by ITIL with the high-level control objectives of COBIT. ITIL was divided into information requirements, in the detailed mapping the information requirements were mapped to COBIT control objectives. The coverage of the mapped information requirements is noted in four different levels: (E) Exceeded, (C) Complete coverage, (A) Some aspects addressed, (N/A) Not addressed.

The management control tools help the manager to translate the strategy to operative terms, because it develops a strategic planning and management with the goal to align business activities with the organization strategy, improves internal and external communications, and monitors organization performance against strategic goals in order to continuously improve strategic performance and results [26].

The main idea is to define a BSC cascade according with the elements expressed above. To define the indicators for each BSC level we propose to use the aggregation methods taking into account: the indicators defined in the COBIT and ITIL framework as showed in the Fig. 3 and will also be a good practice to use the low level indicators in the BSC cascade as source to define more global indicators in the high level for the BSC cascade. The methodology used to define new indicators is described in Sect. 3.1.

Once we have defined the indicators and the relationships among them, then we can design a BI solution that supports the necessary information for the decision making process on the organization.

The first step of the Data Management and sometime the most difficult and hard is the extract, transform, and load (ETL) process, because the information is distributed throughout the length and wide of organizational systems, with different technology and different models. To have a BSC cascade will help us to know:

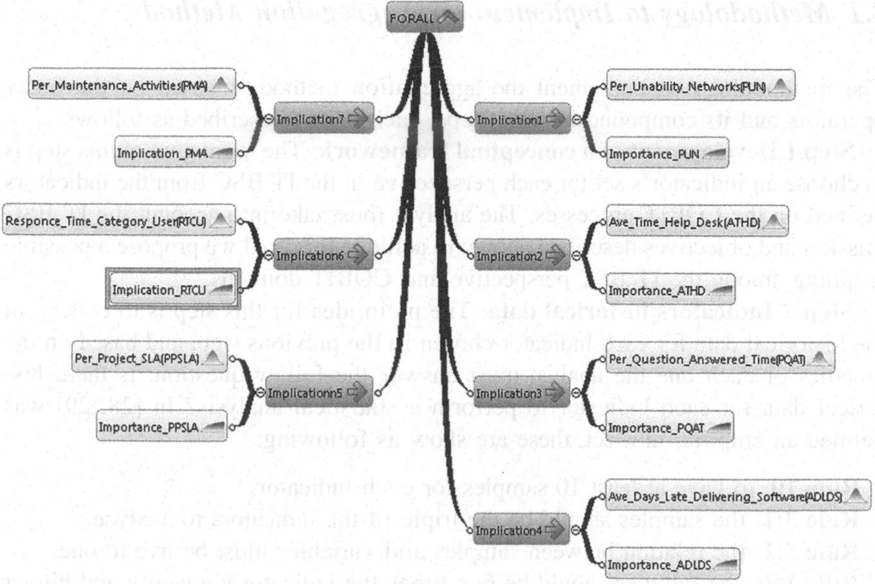

Fig. 3 Fuzzy tree to calculate a global indicator

1. What will be the necessary information to load in the data warehouse?
2. What will be the necessary changes to make on the data before loading them in the data warehouse?
3. Which are the technologies that support the necessary data to load in the data warehouse?
4. Where is distributed the necessary data to load in the data warehouse?
5. When will it be necessary to load the information in the data warehouse?
6. Which will be the main roles of BI?
7. Which will be the necessary reports?

To know the answer of these questions before developing BI is very important to reduce cost and time. Sometimes the cost and the time increase because the company staff does not know the answer of these questions and then they lost a lot time developing patches to fix errors previously committed. The problem is that to develop BI successfully is necessary first an alignment between business domain and technology domain, and is also necessary a common understanding among the business staff and the technological staff, because BI is: "An interactive process for exploring and analyzing structured, domain-specific information (often stored in data warehouses) to discern business trends or patterns, thereby deriving insights and drawing conclusions. The business intelligence process includes communicating findings and effecting change. Domains include customers, suppliers, products, services and competitors" [27].

3.1 Methodology to Implement an Aggregation Method

The methodology to implement the aggregation method is based on the fuzzy operators and its compound by nine steps, each one is described as follows:

Step 1 Development of a conceptual framework: The main goal of this step is to choose an indicator's set for each perspective in the IT BSC from the indicators defined on the COBIT processes. The analyst must take into account the IT BSC mission and objectives described above; to achieve this goal we propose a possible mapping among the IT BSC perspective and COBIT domains.

Step 2 Indicators historical data: The main idea for this step is to collect all the historical data for each indicator chosen in the previous step, and based on the quantity of each one the analyst must answer the follow question: Is there historical data for each indicator to perform a statistical analysis? In [28, 29] was defined an empiric rules set, these are show as following:

- **Rule 10**: to have at least 10 samples for each indicator.
- **Rule 3:1**: the samples should be the triple of the indicators to analyze.
- **Rule 5:1**: the relation between samples and variables must be five to one.
- **Rule 100**: the samples should be five times the indicator's quantity and higher than 100.
- **Rule 150**: to have more than 150 samples for each indicator when the correlation among the indicators is low.
- **Rule 200**: to have more than 200 samples for each indicator, without taking into account the amount of variables.
- **Rule of the significance**: to have 51 more samples than the indicator's quantity, with the purpose of running the Chi squared test.

Sometimes in the enterprise there is not historical data to develop a statistical analysis for each indicator, in this case the analyst can: (1) exclude the indicator from the analysis but if according with the expert criteria the indicator is very important in the analysis then we must to include it and proceed according to step 3 or (2) use a forecasting method to generate the necessary data (this option is applicable if there is some historical data and is only necessary to create a few data samples to fulfill the previous rules).

Step 3 Imputation of missing data: Sometimes we have missing data in the indicators under study; it can affect the final result of global indicators, in [29] were showed four solutions for this problem:

1. Remove the information: in this case we should hide the item from the analysis, it can increase the dispersion.
2. Delete the indicator from the analysis: in this case we should remove the indicator from the analysis. We must to take into account that if the indicator missing values is less than 5 % of all indicators then we can't remove it from the analysis.

3. Single imputation data: we can use statistical methods like average, median, mode or regression based on the available data.
4. Multiple imputations: in this case we can use the Monte Carlo algorithms and Markov chains.

To achieve this step we can use software like: SPSS, Microsoft Excel, Weka, R-Studio.

Step 4 Principal Component Analysis (PCA): the main goal in this step is to know how we can group the indicators to build a global indicator based on the statistical analysis. The algorithm was created by [30] from a geometric approach and afterwards in [31, 32] was defined in an algebraic approach. The algorithm ensures the maximum correlation among the items in the component and minimum correlation among the components, it is an important remark because sometimes it is necessary to define a weight for each indicator or component according with the organization goals or the experts criteria; and if there is relation among components then the indicators within each component will be overweighed. In this step the analyst should validate that the Correlation Matrix's Determinant should be near to zero and the Kaiser-Meyer-Olk in Measure of Sampling Adequacy should be more than 0.5. In the case that any of this parameters is wrong then we advise to use other extraction methods like: Generalized Least Squares, Maximum Likelihood, Alpha Factoring or Unweighted Least Squares where the previous parameters are acceptable.

To achieve this step we can use software like: SPSS, Weka, R-Studio.

Step 5 Internal consistency test: Once we have defined the possible indicators set in each component, it is important to validate the not existents spurious relation among the indicators into a group, to do it we propose to use Cronbach's alpha test [33]. The final value result from the test should be near to one, but according with [33] if the result is higher than 0.7 then it is acceptable. Also we advise to re-calculate the coefficient excluding one indicator per time and if the coefficient value is higher than the previous one then it means that there is spurious relation and we should remove the indicators from the component.

Step 6 Choose components: the main goal in this step is validate according with the expert criteria if the component (global indicator) is a good measuring to measure the goals accomplish.

Step 7 Indicator fuzzification: Every time we want to build new indicators based on aggregation methods is important to normalize the indicators. Under the principles stated above and using compensatory fuzzy logic as an aggregation method we propose to do it using a sigmoidal membership function, because, by theoretical considerations in [34], it is recommend the use of sigmoidal membership functions for increasing or decreasing functions. The parameters of these functions are determined by setting two values. The first is the value at which it is considered that the statement in the predicate is true (gamma). The second is the value for which the data makes almost unacceptable the corresponding statement (beta) [35]. The sigmoidal membership function is calculated as follows:

$$S(x, \alpha, \gamma)_k = \frac{1}{1 + e^{-\alpha(x_k - \gamma)}} \quad (1)$$

$$S(x, \alpha, \gamma)_k = \frac{1}{1 + e^{-\alpha(x_k - \gamma)}} \quad (2)$$

where:

S: Value of truth of the criterion of measurement of indicator "k"

X: Calculated value of the indicator "k" according to the company

Gamma (γ): Value acceptable. It would be equal to the value at which the indicator is considered acceptable.

Beta (β): Value almost unacceptable: It would be equal to the pre-image of a symmetric sigmoidal function for the optimal value defined for the indicator, or it would be the same β = (Value at which the indicator is acceptable—Value from which the indicator is optimal).

Alfa (α): Sigmoidal function parameter. View expression (2).

Step 8 Indicator's importance by component: Sometimes the indicators have different weights according with the organization's goals, for this reason in this step the expert should define a weight for each indicator by global indicator. The scale to define the weighs will be continuous number among [0, 1].

Step 9 Aggregation fuzzy methods: A global indicator is build taking into account the weight for each simple indicator and its value of truth (see expression 1). Under the principles stated above and using compensatory fuzzy logic to compensate the global indicator, would be defined as follows:

$$GI_i = \forall_{j=1}^{j=n}[W_j \rightarrow V_j] \in [0, 1] \quad (3)$$

where:

GI_i: Value of truth of the global indicator "i"

W_j: Weight of the "j" simple indicator

V_j: Value of truth of the simple indicator "j". See expression (1).

The result set from the expression (3) will be a continuo's number among [0, 1] where one is the optimal result and zero the worst result.

4 Partial Results

The performance of the proposed method and its algorithms are evaluated using a set of 20 KPI. The Table 2 shows how with three global indicators we explain near to the 80 % of the total variance. The Cronbach's alpha test is show in the Table 3, and the result is greater than 0.7. According with the result we model the global indicator in the Fuzzy Tree Studio Software (see Fig. 3).

Once defined the indicators for each BSC, we can design the ETL process and also define the OLAP, ad hoc and standard reports.

Table 2 Total variance explained

Component	Initial eigenvalues			Extraction sums of squared loadings		
	Total	% of Variance	Cumulative (%)	Total	% of Variance	Cumulative (%)
1	8.186	38.979	38.979	8.186	38.979	38.79
2	5.982	28.487	67.466	5.982	28.487	67.466
3	2.569	12.233	79.700	2.569	12.233	79.700
4	0.957	4.557	84.257			
5	0.899	4.282	88.539			
6	0.674	3.210	91.749			
7	0.463	2.203	93.952			
8	0.309	1.473	95.426			
9	0.263	1.252	96.678			
10	0.228	1.086	97.763			
11	0.175	0.834	98.598			
12	0.116	0.551	99.148			
13	0.082	0.389	99.537			
14	0.041	0.196	99.733			
15	0.022	0.105	99.839			
16	0.020	0.096	99.935			
17	0.013	0.061	99.996			
18	0.000	0.002	99.998			
19	0.000	0.001	99.999			
20	0.000	0.000	100.000			

Table 3 Reliability statistics

Cronbach's alpha	Cronbach's alpha based on standardized items	N of items
0.846	0.998	7

Item-total statistics

	Scale mean if item deleted	Scale variance if item deleted	Corrected item-total correlation	Squared multiple correlation	Cronbach's alpha if item deleted
Percentage unavailability of network	94561.59854	3.406E11	0.997	0.994	0.867
Average answer time of help desk	94854.26788	3.431E11	0.959	0.947	0.870
Percentage of questions answered within time	65430.59557	1.681E11	1.000	1.000	0.747
Average days late in delivering software	89758.93064	3.111E11	1.000	1.000	0.839
Percentage of projects performed within sla	69306.59726	1.678E11	1.000	1.000	0.747
Percentage of maintenance activities	65619.06807	1.680E11	1.000	1.000	0.747
Response times per category of users	89760.79859	3.111E11	1.000	1.000	0.839

5 Conclusion and Future Work

In the current research we presented some early stage work in a new architecture for BI design. This architecture allows to define the necessary information for the decision making process. The methodology to implement a fuzzy aggregation method allows us to define new global indicators supported in statistical and fuzzy approach. As future work we propose to make a mapping among the COBIT domains and the IT BSC.

References

1. Lucio-Nieto, T., et al.: Implementing an IT service information management framework: the case of COTEMAR. Int. J. Inf. Manage. **32**, 589–594 (2012)
2. Valiente, M.-C., Garcia-Barriocanal, E., Sicilia, M.-A.: Applying an ontology approach to IT service management for business-IT integration. Knowl. Based Syst. **28**, 76–87 (2012)
3. ITGI: COBIT4.1. Control objectives for information and related technology. http://www.itgi.org/COBIT.htm (2007)
4. ITIL: Information technology infrastructure library V.3. http://www.itil.co.uk (2007)
5. Espin Andrade, R.A., et al.: Compensatory logic: a fuzzy formative model for decision making. In: Congress of International Association for Fuzzy-Set Management and Economy, León, España (2003)
6. Gold, C.: Total Quality Management in Information Services – IS Measures: A Balancing Act. Research Note Ernst and Young Center for Information Technology and Strategy, Boston (1992)
7. Willcocks, L.: Information Management. The Evaluation of Information Systems Investments. Chapman and Hall, London (1995)
8. Van Grembergen, W., Van Bruggen, R.: Measuring and improving corporate information technology through the balanced scorecard technique. In: Fourth European Conference on the Evaluation Of Information Technology, Deflt (1997)
9. Van Grembergen, W., Timmerman, D.: Monitoring the IT process through the balanced scorecard. In: 9th Information Resources Management (IRMA) International Conference, Boston (1998)
10. Van Grembergen, W.: The balanced scorecard and IT governance. Inf.Syst. Control J. **2**, 40–43 (2000)
11. Van Grembergen, W., De Haes, S.: The IT balanced scorecard as a framework for enterprise governance of IT. In: Enterprise governance of information technology. Achieving strategic alignment and value. Springer, New York (2009)
12. Van der Zee, J.: Alignment is not enough: integrating business and IT management with the balanced scorecard. In: 1st conference on the IT balanced scorecard, Antwerp (1999)
13. Van Grembergen, W., Saull, R., De Haes, S.: Linking the IT balanced scorecard to the business objectives at a major canadian financial group
14. Fatemeh Moghimi, C.Z.: A decision-making model to choose business intelligence platforms for organizations. In: Third International symposium on intelligent information technology application, IEEE (2009)
15. Ying Wang, Z.L.: Study on port business intelligence system combined with business performance management. In: Second international conference on future information technology and management engineering, IEEE (2009)

16. Merigó, J.M., Gil-Lafuente, A.M.: Fuzzy induced generalized aggregation operators and its application in multi-person decision making. Expert Syst. Appl. **38**(8), 9761–9772 (2011)
17. Wei, G., Zhao, X.: Some induced correlated aggregating operators with intuitionistic fuzzy information and their application to multiple attribute group decision making. Expert Syst. Appl. **39**(2), 2026–2034 (2012)
18. Zandi, F., Tavana, M.: A fuzzy group multi-criteria enterprise architecture framework selection model. Expert Syst. Appl. **39**(1), 1165–1173 (2012)
19. Cebeci, U.: Fuzzy AHP-based decision support system for selecting ERP systems in textile industry by using balanced scorecard. Expert Syst. Appl. **36**(5), 8900–8909 (2009)
20. Yüksel, I., Dağdeviren, M.: Using the fuzzy analytic network process (ANP) for balanced scorecard (BSC): a case study for a manufacturing firm. Expert Syst. Appl. **37**(2), 1270–1278 (2010)
21. Grabisch, M., et al.: Aggregation functions: construction methods, conjunctive, disjunctive and mixed classes. Inf. Sci. **181**(1), 23–43 (2011)
22. Fields, E.B., Okudan, G.E., Ashour, O.M.: Rank aggregation methods comparison: a case for triage prioritization. Expert Syst. Appl. **40**(4), 1305–1311 (2013)
23. Ting, A.: C omparison of different aggregation methods in coupling of the numerical precipitation forecasting and hydrological forecasting. Procedia Eng. **28**, 786–790 (2012)
24. Tsyganok, V.: Investigation of the aggregation effectiveness of expert estimates obtained by the pairwise comparison method. Math. Comput. Model. **52**(3–4), 538–544 (2010)
25. ITGI: COBIT MAPPING: Mapping of ITIL with COBIT 4.1. ITGI (2008)
26. Balanced Scorecard Institute: What is the balanced scorecard? http://www.balancedscorecard.org/BSCResources/AbouttheBalancedScorecard/tabid/55/Default.aspx (2010) Accessed 04 Aug 2010
27. Group, G.: The gartner glossary of information technology acronyms and terms. http://www.gartner.com/6_help/glossary/Gartner_IT_Glossary.pdf (2004)
28. Nardo, M., et al.: Handbook on constructing composite indicators: methodology and user guide. OECD Statistics Working Paper, STD/DOC, Paris (2005)
29. Little, R., Rubin, D.: Statistical analysis with missing data. John Wiley, New York (2002)
30. Pearson, K.: On lines and planes of closest fit to a system of points in space. Philos. Mag. **2**, 559–572 (1901)
31. Hotelling, H.: Analysis of a complex of statistical variables into principal components. J. Edu. Psychol. **24**, 417–441 (1993)
32. Kaiser, H. The varimax criterion for analytic rotation in factor analysis. Psychometrika **23**, 187–200 (1958)
33. Cronbach, L. Coefficient alpha and the internal structure of tests. Psychometrika **16**, 297–334 (1951)
34. Dubois, D., Prade, H. Review of fuzzy set aggregation connectives. Inf. Sci. **36**, 85–121 (1985)
35. Ceruto Cordovés, T, Rosete Suárez, A., Espín Andrade, R. A. Descubrimiento de predicados a través de la búsqueda metaheurística (2009)

Chapter 11
Improving IT Service Management with Decision-Making Support Systems

Manuel Mora, Gloria Phillips-Wren, Francisco Cervantes Pérez, Leonardo Garrido and Ovsei Gelman

Abstract IT Service Management (ITSM) is a managerial approach to deliver value through IT services. This service-oriented world-view has required new knowledge on processes and tools to cope with the planning, design, building, operation, and improvement of IT services. While some decisions can be efficiently and effectively made by a manager alone, more complex decisions require further analysis and can benefit from the use of computerized Decision-Making Support Systems (DMSS). Although DMSS have been available for several decades, they have not been used to support ITSM processes and decisions. The purpose of this chapter is to demonstrate that ITSM would benefit from the utilization of DMSS. To this end, we provide a description of the decision-making process and decisional services embedded in the DMSS architecture and analyze how ITSM processes and decisions can be supported by DMSS. This chapter aims to foster more specific research on DMSS for ITSM.

Keywords ITSM · DSS · DMSS · ITIL

M. Mora (✉)
Autonomous University of Aguascalientes, 20131 Aguascalientes, AGS, Mexico
e-mail: mmora@securenym.net

G. Phillips-Wren
Loyola University, Baltimore, MD 21210, USA
e-mail: gwren@loyola.edu

L. Garrido
ITESM, 64849 Monterrey, NL, Mexico
e-mail: leonardo.garrido@itesm.edu

F. Cervantes Pérez · O. Gelman
CCADET, Universidad Nacional Autónoma de México, 04510 Mexico city, DF, Mexico
e-mail: francisco.cevantes@ccadet.unam.mx

O. Gelman
e-mail: ogelman@unam.mx

M. Mora et al. (eds.), *Engineering and Management of IT-based Service Systems*,
Intelligent Systems Reference Library 55, DOI: 10.1007/978-3-642-39928-2_11,

1 Introduction

Services, from a business view, have changed in their economic impact over the last 30 years [1]. Nowadays, business organizations focused on delivering "*help, utility, experience, information or other intellectual content… account for more than 70 % of total value added in the OECD*" [2]. This economic perspective shift from product design, product manufacturing, and product distribution to service design, service composition, and service delivering, can be explained from a market focus on guaranteed functionalities of systems as products per se or as systems using products, process, technology and people [3]. Services concept is not unique. Several conceptualizations have been posed primarily in business and operations management literature [4–6]. Quinn [4, p. 5] describes services as outputs of a service sector business where service: "*is not a product or construction, is generally consumed at the time it is produced, and provides added value in forms (such as convenience, amusement, timeliness, comfort, or health) that are essentially intangible concerns of its purchaser.*" Spohrer et al. [5, p. 72] provides a similar description as: "*Service is the application of competences for the benefit of another, meaning that service is a kind of action, performance, or promise that's exchanged for value between provider and client.*" In Mora et al. [6, p. 17] these ideas are extended to define service as a 3-dimensional concept that consists of interactions between service facilitator and service appraiser, changes in attributes of interest in service facilitator and service appraiser, and co-value generation in the joint service facilitator-appraiser system.

Service emerges as fundamental in the Information Technology (IT) area given: (i) the complexity of managing business processes daily, tactically and strategically in today's global organizations [7]; and, (ii) the large IT-investments that are used to address these business complexities [8]. Necessarily, organizations now rely strongly on complex, IT-based systems [9] which are responsible for delivering sophisticated IT services. Although an IT service concept has been incorporated previously in the IT literature, its study has been implicit [10] as any action received from the IT area toward IT users. An updated concept of service [5] is being used in current ITSM research [11] as a fundamental IT concept that embraces the modern business concept of service. Consequently, several ITSM process frameworks for engineering and management IT services have been proposed: ISO/IEC 20000 [12, 13], ITIL v3 [14], CMMI-SVC [15], ITUP® [16], and MOF® 4.0 [17]. More specifically, IT Service Management (ITSM) can be defined as a management system of organizational resources and capabilities for providing value to organizational customers through IT services [14]. ITSM has become a relevant issue for large and mid-sized organizations because it is expected that IT services, together with IT architecture, will deliver more efficient and effective IT management, and ultimately better organizational value [18]. In particular, in [18] is identified the need of enterprises to rely more on IT-supported business processes and IT services as the vehicle for delivering value. Studies on ITSM impacts are still scarce, but some empirical evidence of organizational benefits from ITSM has been reported [19]. For example, in [19] in a survey of 65 Australian corporations, were identified the following benefits: improved customer satisfaction, improved response and resolution time, improved IT service continuity, clear identification of roles/responsibilities, reduction in cost/incident, and improved IT employee productivity related to ITSM.

In this chapter, we aim to advance knowledge on the application of ITSM process frameworks for both theory and practice. We also suggest that the application of ITSM process concepts will help practitioners in simple, moderate and complex decisional situations. A decisional situation occurs when at least two courses of action are feasible but it is not clear which of them should be selected. The process to analyze such situations is called a decision-making process (DMP). DMP have been considered one of the most critical and central activities performed in organizations [20]. Several frameworks have been reported to support the DMP [21]. In particular, decision makers do not typically require support for simple decisions (e.g., few courses of action, clear distinction of their impacts on the criterion of interest, and a low economic impact of mistakes). However, for moderate and complex decisions, decision makers have used structured DMP and special computer-based support tools called collectively Decision-Making Support Systems (DMSS) [22] over the last three decades. Such DMSS can be enabled with analytical capabilities or enhanced with intelligent mechanisms.

Unexpectedly, a review of ITSM literature reveals that DMSS research is scarce in this domain. This gap in the literature presents an opportunity to examine the case for fundamental and applied research on how DMSS can help ITSM practitioners apply ITSM process frameworks. Consequently, in this chapter we present a review on the decision making process (DMP) and DMSS in Sect. 2. Then we examine the ITSM process framework and related ITSM decisions in Sect. 3. An analysis and recommendations on how DMSS (and, in particular, intelligent DMSS) can support such ITSM decisions is presented in Sect. 4. Conclusions are presented finally in Sect. 5. Through this chapter, we aim to advance theoretical and practitioner knowledge on the enhancement of ITSM process utilization through the incorporation of DMSS and intelligent DMSS.

2 Decision-Making Process and Intelligent Decision-Making Support Systems

2.1 Decision-Making Process

A Decision-making Process (DMP) can defined as "… *the sensing, exploration and definition of problems and opportunities as well as the generation, evaluation and selection of solutions*" and *the implementation and control of the actions required to support the course of action selected* [20, 23]. A systematic model [24] can guide decision makers on the DMP. Simon's DMP [24] consists of three phases: (i) "Intelligence" where the decision maker identifies and prioritizes an organizational problem which requires a solution and goals to be reached; (ii) "Design" where a model to structure the decision situation is formulated, action alternatives (i.e. the set of possible decisions) and their associated outcomes are generated and estimated, and the evaluation criteria are established; and (iii) "Choice" where the decision maker evaluates the outcomes of each alternative and selects the action that best achieves the decision objectives.

DMP model extensions have been proposed based on Simon's model [25]. In this chapter, we describe a generic DMP model that includes the main aspects reported in previous DMP models [21]. As suggested by Simon [24], the DMP Model (see Fig. 1) treats the decision steps within each phase as iterative rather than sequential.

This generic DMP has five main phases. In the first "Intelligence" phase, the initial step is "Problem Detection", which has the purpose of detecting a potential problematic situation requiring a decision-making action. A decision problem is any relevant and non-desired or improvement-based organizational situation which demands an agreement on what corrective or improvement organizational actions must be deployed. While potential actions may be known in advance, the best decision based on the criteria is not known. "Data Gathering" is performed with "Problem Formulation" to collect data and organize relevant information for establishing a "Problem Scheme". This first phase ends when the "Problem Scheme" deliverable is generated and agreed upon by the decision maker or team. The "Problem Scheme" usually includes problem symptoms, problem owners, economic problem relevance, problem statement, and decisional objectives.

The "Design" phase uses the "Problem Scheme" as the key input for the "Model Classification" step. The decision maker or team (including any advisory staff) analyze the elements reported in the "Problem Scheme" for mapping into either an already known organizational "Model Scheme" or for elaborating a new one (usually an approximate one for the current "Problem Scheme"). A "Model

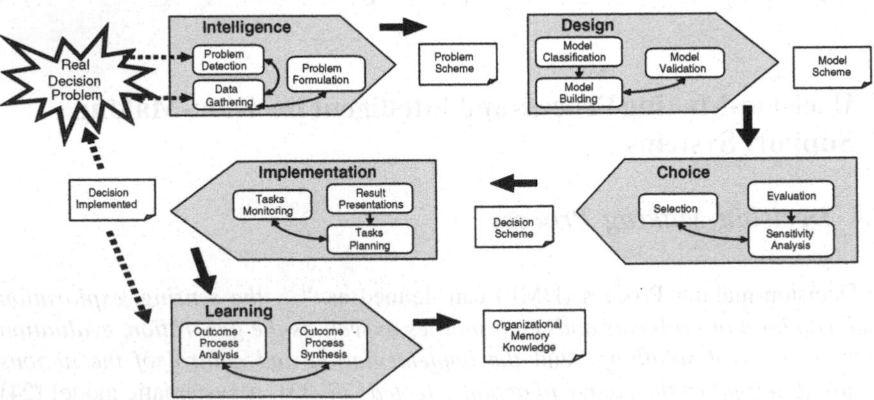

Fig. 1 Generalized decision-making process model

Scheme" contains at least: a set of courses of action (X variables), a set of known or controlled parameters (W variables), a set of outputs (Y variables), a set of uncertain or risk-based, non-controlled environmental items (Z variables), a set of behavioral interrelationships between such X, W, Y and Z variables (with underlying analytic or stochastic functions), a hierarchy of criteria with their evaluation scales, respective values and associated risk preferences.

The decision maker or team uses their knowledge to develop a "Model Scheme" which suitably fits the "Problem Scheme". A particular case for the "Model Scheme" is instanced using the current data provided by the "Problem Scheme" in the "Model Building" step. Most required data can be used from previous cases, although the decision maker or team is free to assign new values for particular variables. If the decisional problem requires a new "Model Scheme", the organization can either elaborate an approximate model with internal expertise or ask for an outsourced design with external consultants. In both cases, this phase ends with a "Model Validation" step where face validity tests are performed by the decision maker or team to validate the model, data and interrelationships between variables in the "Model Scheme". Thus, the adequacy and correctness of the model and its components are evaluated. At the end of this phase, the decision maker will have a "Model Scheme" suitable to be solved computationally.

The third phase of "Choice" is composed of three well known steps: "Evaluation", "Sensitivity Analysis" and "Selection". In the "Evaluation" step, the "Model Scheme" is solved or resolve. Solving the model implies obtaining an optimized solution when it is feasible. Resolving the model means that a non-optimal, but satisfactory, solution is obtained. In both cases a solution is a ranked list of courses of action according to the hierarchy of criteria. "Sensitivity Analysis" is a required step for identifying very sensitive or very robust items in the "Model Scheme". A sensitive item is one which shows large changes with small changes in another variable or variables in the model. In contrast a robust item is one which has small changes despite large changes in another variable or variables. The "Sensitivity Analysis" step is useful to break ties in the solution set, to create awareness for decision-makers on potential large changes in some variables of interest, and to add an additional evaluation criterion: the robustness of the preferred course of action. In the third step of "Selection" the decision maker or team chooses the preferred course of action based on the results provided by the previous two steps. At the end of this phase, the decision maker has a "Decision Scheme" that identifies the best alternatives based on the established criteria and the "Sensitivity Analysis".

The "Implementation" phase consists of three steps: "Result Presentation", "Task Planning" and "Task Monitoring". During "Result Presentation" the decision or the set of best decisions are communicated to the executive team for authorization. Next, in "Task Planning", the decision is translated into an operational and scheduled plan of tasks, resources, deliverables and performance metrics. Finally, in "Task Monitoring", the execution, control and monitoring of the actions derived from the decision are carried out. The last phase in the DMP model is called "Learning". Two steps are included in this phase:

"Outcome-Process Analysis" and "Outcome-Process Synthesis". In the "Outcome-Process Analysis" step, several metrics of process and outcome efficiency, efficacy and effectiveness is collected from decision-makers. During the "Outcome-Process Synthesis" step, key lessons learned are discussed and holistically synthesized
by the decisional team under a philosophy of continuous improvement and organizational learning. New knowledge resulting from both steps should be incorporated into organizational memory.

2.2 *Intelligent Decision-Making Support System Architecture*

The Integrated Design and Evaluation Framework for i-DMSS (IDEF-i-DMSS) [21, 26] is a framework that integrates the needs and findings of the DMSS literature with advances in artificial intelligence (AI) for the purpose of improving the design of i-DMSS. In the IDEF-i-DMSS, well-known knowledge and computational representation levels used in AI systems are used in four levels to capture a comprehensive design and evaluation view. These extensions link decision-making phases and steps with decisional services/tasks, architectural capabilities, and computational symbol/program mechanisms. The main theoretical premise of the IDEF-i-DMSS framework is: *decision-making phases and steps can be improved by the support of decisional services and tasks, which are provided by architectural*

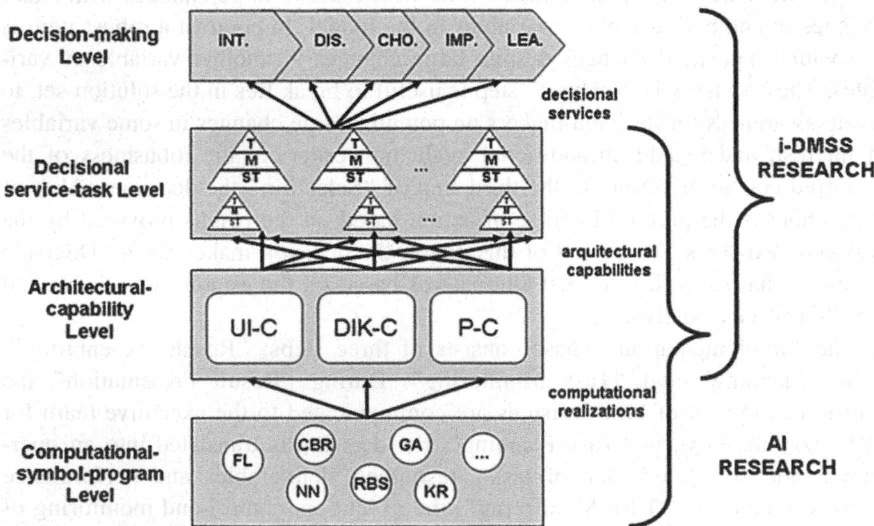

Fig. 2 Comprehensive framework for design and evaluation of i-DMSS

capabilities that can/could in the future be computationally implemented by symbol/program-based mechanisms. Figure 2 shows this framework.

The top level, the Decision-making Level, includes the decision-making phases and activities to be executed by a decision-maker using an i-DMSS. The second level, the Decisional Service-task Level, accounts for the decisional support services of the i-DMSS (e.g. the Newell's Knowledge Level required in AI systems). The third level, the Architectural-capability Level, includes the user interface, data-information-knowledge and processing capabilities provided by the components of the architecture of the i-DMSS. Finally, the lowest level, the Computational/Program/Symbol Level, includes the specific AI computational mechanisms that implement the architectural components of the i-DMSS.

The phases and steps included in the Decision-making Level have been reported earlier in the DMP model. These DMP model phases are: Intelligence, Design, Choice, Implementation and Learning. The steps within the phases are: Problem Detection, Data Gathering, Problem Formulation, Model Classification, Model Building, Model Validation, Evaluation, Sensitivity Analysis, Selection, Result Presentations, Task Planning, Task Monitoring, Outcome-Process Analysis and Outcome-Process Synthesis. The DMP model's phases and steps are expected to be supported by the decisional services provisioned by the second level in an i-DMSS.

In the Service-task Level there are a set of generic tasks (algorithmic mechanisms, heuristic analysis, heuristic synthesis or hybrid ones) and generic services: find, alert, optimize, classify, monitor, interpret, predict, configure, schedule, formulate, plan, explain, recommend, modify, control, learn, and instruct. Table 1 reports these decisional services and tasks at a high-level of abstraction. The development strategy is to consider decisional services as generic building-blocks to design an i-DMSS. An actual i-DMSS will provide only those tasks and services needed to perform its function.

The next level in Fig. 2, the Architectural-capability level, is an intermediate level between the knowledge level (e.g. the decisional service-task level) and the computational level. This intermediate level considers three dimensions: (i) user interface processes, (ii) data, information and knowledge representation, and (iii) support processing. The first and second dimensions are based on the general and standard structure for a DMSS [27]. The third dimension is based on the types of decisions tasks, levels of intelligence embedded in algorithms and types of intelligent operations for intelligent data mining systems suggested in the literature [28, 29]. Tables 2, 3 and 4 present updated descriptions of these dimensions.

These tables represent ordinal conceptual scales to measure the degree of: (i) user interface capability, (ii) rawness in the data component and (ii) the degree of intelligence embedded in the algorithms or processing mechanisms of a particular i-DMSS. It must be noted that any support level includes, or can include, capabilities from a previous level. Finally, in the lowest level, the Computational/Program/Symbol Level, specific AI-based and non-AI mechanisms are used to enable the architectural capabilities of the next upper level.

We propose that this framework offers a more detailed conceptual scheme than previous ones reported in the literature. We further propose that this framework is

Table 1 Taxonomy of generic decisional services and tasks

Task type	Generic services (inputs): outputs	Generic tasks
Algorithmic mechanisms	Find(query, system): result-set	Retrieval
	Alert(conditions, system): result-set	Triggering
	Apply-madm(decision-data): decision-result	Calculation
	What-if(variable-set, original-model): modified-model	Sensitivity analysis
	Goal-seeking(goal-variable, original-model): modified-model	Sensitivity analysis
	Identify-critical-variables(variable-set, original-model): critical-variable-set	Sensitivity analysis
	Maxormin(goals, constraints, model): result-set	Optimization
Heuristic analysis	Classify(data, system): system-pattern	Classification
	Monitor(system, metrics-set): (system-variations, causal-links of variations)	Classification
	Interpret(data, system): system-state-assessment	Identification
	Predict(system, events-set, time-period): future-system-state	Identification
Heuristic synthesis	Configure(parts, constrains, goals): system-structure	Design
	Plan-schedule(activities, resources, constrains, goals): (states-sequence, system-structure) states-sequence	Design
	Formulate-design(components, goals, constrains): system-structure	Complex design
Heuristic hybrid	Explain(data, system): system-cause-effect-links	Complex
	Recommend(base system, required system): change-actions	Complex
	Control(system-state, goals):input-system-actions	Complex
	Discover(data, system): knowledge-structures	Complex
	Learn (system, knowledge-on-system): new-knowledge	Complex

Table 2 Levels of user interface capability dimension

User interface levels	Description
I. Text and graphics	A DMSS uses as action language structured commands and/or menus and as presentation language texts and graphics
II. Multimedia	A DMSS uses as action language structured commands and/or menus and as presentation language texts, graphics, sound, animations and video
III. Advanced UI	A DMSS uses as action language natural plain language and as presentation language all previous issues enhanced by virtual reality environments

useful to analyze what decisional support capabilities can be provisioned in an i-DMSS for ITSM tasks with a greater level of conceptual detail.

Table 3 Levels of data, information and knowledge representation capability dimension

Data, information and knowledge levels	Description
I. Data bases	A DMSS uses plain files, simple data structures or/and relational databases scheme to represent data and information
II Multidimensional data bases	A DMSS uses complex and highly structured data structures or/and multidimensional databases schemes to represent data and information
III. Numerical models	A DMSS accesses structured data, information and knowledge organized in numerical models, such as forecasting models, simulation models, statistical models, Bayesian networks, and neural layers
IV. Knowledge bases	A DMSS accesses highly semi-structured data, information and knowledge organized in knowledge chunks. Examples of these schemes are semantic networks, rules, fuzzy rules, frames, scripts, and cases
V. Distributed knowledge bases	A DMSS accesses a network of highly ill-structured data, information and knowledge organized in knowledge bases. Examples of these schemes are ontological knowledge repositories

Table 4 Levels of processing capability dimension

Processing levels	Description
I. SQL-based	A DMSS supports the all SQL actions: searching, adding, updating, deleting and sorting using a crisp logic mechanism. Also it supports drilling-drown, rolling-up, slicing and pivoting operations for multi-dimensional data warehouses
II. MADM and optimization	A DMSS supports operations of optimization, classification, association, clustering, trend analysis and forecasting where problems are intensive on quantitative or numerical-based data. Examples of processing mechanisms are: MCDA algorithms, neural networks, genetic algorithms, data mining and statistical-based algorithms
III. Fuzzy SQL-based	A DMSS supports the all SQL-based actions: searching, adding, updating, deleting and sorting using a fuzzy logic mechanism. Also it supports drilling-drown, rolling-up, slicing and pivoting operations for messy data
IV. Semi-structured problem solving	A DMSS supports intelligent algorithms for complex analysis tasks such as: classification, diagnosis, interpretation and monitoring/control. Examples are: rule-based inference algorithms, case-based techniques and frame and semantic networks inference algorithms
V. Ill-Structured problem solving	A DMSS supports intelligent algorithms for complex synthesis tasks such as: exploring, explanation, planning, design and learning. Examples are: agent-based algorithms and natural-language processing mechanisms

3 IT Service Management Processes-Decisions

3.1 ITSM Process Framework

IT Service Management (ITSM) was presented previously as a management system of organizational resources and capabilities for providing value to organizational customers through IT services [14]. While several ITSM process frameworks for engineering and management IT services have been proposed (ISO/IEC 20000 [12, 13], ITIL v3 [14], CMMI-SVC [15], ITUP® [16], and MOF® 4.0 [17]), the ITIL v3 [14] process framework is one of the most accepted in practice. ITIL v.3:2007 provides "*a common framework of practices that unite all areas of IT service provision toward a single aim – delivering value to the business*" [14, p.3]. Value is delivered through the accomplishment of the expected or agreed upon warranty and the co-generated utility of the provisioned IT services. In this way, services must be considered as business assets with two core attributes: utility (fit for purpose) and warranty (fit for use). The greater the positive effects on the task performance related to desired customer and business objectives, the greater the utility perceived by the customer. Similarly, the greater the availability, capacity, continuity and security of service provision, the greater the warranty [14].

Consequently, ITIL v.3:2007's overall objective is "*to provide services to business customers that are fit for purpose, stable and that are so reliable, the business views them as a trusted utility*". For ITIL v3 [14, p.5] services, in general, are "*means for delivering value to customers by facilitating outcomes customers want to achieve without the ownership of specific costs and risks*". In turn, OGC [14] defines an IT service as "*a service provided to one or more customers by an IT service provider, based on the use of IT and supports the customer's business processes, and is made up from a combination of people, processes and technology and defined in a service level agreement.*" The ITIL v3 [14] process framework is structured into five main phases: Service Strategy, Service Design, Service Transition, Service Operation and Service Continual Improvement. Each phase is comprised of processes, functions and additional practices. Table 5 reports such a macro-structure.

Space limitations preclude a detailed review of each process, function and additional practice of ITIL v3. However, the information in Table 5 is useful to estimate the complexity of decisions that ITIL v3 implementers face during the execution of such processes, functions and practices. We illustrate typical decisions for the first two phases in next section.

3.2 Decisions in ITSM Processes, Functions and Practices

To illustrate how DMSS and intelligent DMSS can help ITSM implementers to face moderate and complex decisions, we need to elaborate a plausible catalogue

Table 5 Core process structure of ITIL v3

Phases	Purposes, processes, functions and additional practices
Service strategy	Its purpose is to plan and apply a service strategy aligned with the business strategy. It defines the market space of IT service customers and their associated IT service portfolio, including financial and demand issues in such plan. Its processes are: Strategy Generation, Financial Management, Demand Management, and Service Portfolio Management. Its additional practices are: Return on Investment of IT Services, and Risk Management of IT Services
Service design	Its purpose is "to design appropriate and innovative IT services, including their architectures, processes, policies and documentation, to meet current and future agreed business requirements". Its processes are: Service Catalogue Management, Service Level Management, Capacity Management, Availability Management, IT Service Continuity Management, Information Security Management, and Supplier Management. Its additional practices are: designing supporting systems, designing technology architectures, designing processes, and designing measurement systems and metrics
Service transition	Its purpose is "to deliver services that are required by the business into operational use". Its processes are: Change Management, Service Asset and Configuration Management, Knowledge Management, Transition Planning and Support, Release and Deployment Management, Service Validation and Testing, and Evaluation. Its additional practices are: managing communications and commitment, managing organization and stakeholder change, and managing stakeholders
Service operation	Its purpose is "to deliver agreed levels of service to users and customers, and to manage the applications, technology and infrastructure that support delivery of the services". Its processes are: Event Management, Incident Management, Request Fulfillment Management, Access Management, and Problem Management. Its functions are: Service Desks, Technical Management, IT Operations Management, and Application Management. Its additional practices are: monitoring and control, and IT operations
Continual service improvement	Its purpose is "to keep value for customers through the continual evaluation and improvement of the quality of services and the overall maturity of the ITSM service lifecycle and underlying processes". Its processes are: 7-Step Improvement, Service Measurement, and Service Reporting. Its additional practices are: Deming Cycle, Benchmarking Mechanisms, and Governance Schemes

of usual decisions occurring in the first two ITSM phases (e.g. in a process, function or practice). Tables 6 and 7 report such information.

While this initial catalogue of ITSM decisions is not exhaustive, it is sufficiently illustrative to demonstrate the need to incorporate DMSS and intelligent DMSS tools for supporting such processes, functions and practices. As mentioned previously, a literature review of DMSS and intelligent DMSS applications in ITSM processes revealed little research in this area.

Table 6 Usual decisions in service strategy phase

Itsm process, function or pracice description	Examples of usual decisions with [moderate or complex decision] level
Strategy generation process. ITSM implementers need to define the service market, develop the offerings, develop the strategic assets (capabilities and resources), and prepare execution	1. To assign priority to a set of alternative service offerings regarding their tag customer expected outcomes (e.g. enhance capabilities, increase performance, enhance resources, reduce costs, and reduce risks). [complex decision] 2. To evaluate a set of organizational assets (resources and capabilities) for a better service potential improvement. [moderate decision] 3. To select best service strategies of a set of alternatives regarding expected costs, risks, benefits and release timelines. [complex decision]
Financial management process. ITSM implementers need to define budgeting, accounting, and charging mechanisms	1. To evaluate economically service provision models. [moderate decision] 2. To prioritize investments and budgets demanded for services. [moderate decision] 3. To evaluate economically the value of provisioned services. [complex decision]
Demand management process. ITSM implementers need to identify patterns of business activity, user profiles, and service level packages	1. To rank the relevance of patterns of business activity [moderate decision] 2. To select a reduced number of user profiles. [moderate decision] 3. To select service level packages to be offered. [moderate decision]
Service portfolio management process. ITSM implementers need to conform an IT service portfolio	1. To maximize the expected value of the IT service portfolio. [complex decision] 2. To justify the authorization or rejection of IT services comparing with a minimal expected overall value and cost-benefit of a standard IT service. [moderate decision] 3. To evaluate the introduction of new or enhanced services. [complex decision]
Return on investment of IT Services Practice. ITSM implementers need to calculate or estimate the ROI of current or planned IT services	1. To rank current or planned IT services based on the achieved or estimated ROIs. [moderate decision]
Risk management of IT services practice. ITSM implementers need to evaluate risk on IT services as well as their alternatives of treatment	1. To rank risk exposition of critical of a set of IT services. [complex decision] 2. To evaluate the costs of risk treatments in a set of IT services. [moderate decision]

Table 7 Usual decisions in service design phase

Itsm process, function or pracice description	Examples of usual decisions with [moderate or complex decision] level
Service catalogue management process. ITSM implementers need to select the structure of the business and technical IT services catalogue, and the support tools for managing it	1. To select the structure for business and technical catalogues from several alternatives. [moderate decision] 2. To evaluate commercial and open source tools to be used in SCM. [moderate decision]
Service level management process. ITSM implementers need to negotiate suitable SLAs, OLAs and UCs	1. To select best options from a set of alternative SLAs jointly with users. [moderate decision] 2. To select best options from a set of alternative OLAs for supporting a particular SLA. [moderate decision] 3. To select best options from a set of alternative UCs for supporting a particular SLA. [complex decision]
Capacity management process. ITSM implementers need to select best architecture of CIs in designed IT services for achieving the expected capacities levels	1. To select best architectures of CIs from a set of alternative schemes for normal services. [moderate decision] 2. To select best architectures of CIs from a set of alternative schemes for critical services. [complex decision] 3. To select a particular and critical CI for supporting a critical IT service. [complex decision] 4. To evaluate several schemes for IT teams for achieving the expected levels of service. [moderate decision]
Availability management process. ITSM implementers need to select best architecture of CIs in designed IT services for achieving the expected availabilities levels	5. To evaluate options for external (on the cloud) extra IT capabilities for achieving the expected levels of service. [complex decision]
IT service continuity management process. ITSM implementers need to select best architecture of CIs in designed IT services for achieving the expected continuity levels	
Information security management process. ITSM implementers need to select best information security policies	1. To select best options for information security policies to be applied [complex decision] 2. To rank information assets by relevance regarding security issues. [moderate decision] 3. To evaluate impact of hypothetical disasters on information assets. [moderate decision]

(continued)

Table 7 (continued)

Itsm process, function or pracice description	Examples of usual decisions with [moderate or complex decision] level
Supplier management process. ITSM implementers need to select best IT suppliers for IT assets or as external service providers under UCs or SLAs	1. To select best IT suppliers for acquiring normal IT assets [moderate decision] 2. To select best IT suppliers for acquiring critical IT assets [complex decision] 3. To evaluate the overall performance of IT suppliers. [moderate decision]

4 How DMSS and i-DMSS can Support ITSM Processes and Decisions

In this section we present two illustrative cases on how a DMSS or an i-DMSS can support an ITSM process-decision. Tables 8 and 9 report such cases. In these tables, we use the following structure: (i) column 1 reports the three main phases of the generic DMP reported in Sect. 2; (ii) column 2 reports the particular decisional step; (iii) columns 3 to 5 report the processing, data-information-knowledge, and user interface capability levels of the IDEF-i-DMSS framework [21] reported in Sect. 2; (iv) column 6 reports the particular support (e.g. a decisional service) that an DMSS or intelligent DMSS can provide; and (v) each cell marked with a black dot implies the utilization of an architectural capability. A hyphen in cell means that no architectural capability is suggested.

Without loss of generality, we use the first three phases of a generic decision-making process, and the architectural capabilities and decisional service level in such illustrative descriptions. In both cases the first decision reported in the Phase

Table 8 Illustrative case of a service strategy decision supported for a plausible DMSS

DECISION PHASES AND { DECISION STEPS }	(1) SQL-BASED	(2) MADM AND OPT.	(3) FUZZY-SQL-BASED	(4) SEMI-STRUCTURED PS	(5) ILL-STRUCTURED PS	(1) DBMS	(2) M-DBMS	(3) NUMERICAL MODELS	(4) KNOWLEDGE BASES	(5) DISTRIBUTED KB	(1) TEXT & GRAPHICS	(2) MULTIMEIA	(3) ADVANCED UI	APPLIED SERVICES
	PROCESSING CAPABILITY					DATA-INF-KNWDGE CAPABILITY					USER-INTF. CAPABILITY			
INTELLIGENCE { Problem Detection, Data Gathering, Problem Formulation }	•	•	-	-	-	•	-	-	-	-	•	-	-	ALERT(conditions, system):result-set; FIND(query): result-set
DESIGN { Model Classification, Model Building, Model Validation }	-	-	-	-	-	-	-	-	-	-	-	-	-	-
CHOICE { Evaluation, Sensitivity Analysis, Selection }	-	•	-	-	-	-	-	•	-	-	•	-	-	APPLY-MADM(decision-data):decision-result; WHAT-IF(variable-set, original-model):modified-model; GOAL-SEEKING(goal-variable, original-model):modified-model; IDENTIFY-CRITICAL-VARIABLES(variable-set, original-model):critical-variable-set.

Table 9 Illustrative case of a service strategy decision supported for an intelligent DMSS

	PROCESSING CAPABILITY					DATA-INF-KNWDGE CAPABILITY					USER-INTF. CAPABILITY			APPLIED SERVICES
DECISION PHASES AND { DECISION STEPS }	(1) SQL-BASED	(2) MADM AND OPT.	(3) FUZZY-SQL-BASED	(4) SEMI-STRUCTURED PS.	(5) ILL-STRUCTURED PS	(1) DBMS	(2) M-DBMS	(3) NUMERICAL MODELS	(4) KNOWLEDGE BASES	(5) DISTRIBUTED KB	(1) TEXT & GRAPHICS	(2) MULTIMEIA	(3) ADVANCED UI	
INTELLIGENCE { Problem Detection, Data Gathering, Problem Formulation }	•	•	•	-	-	•	•	-	-	-	•	-	-	ALERT(conditions, system):result-set; FIND(query): result-set; *DISCOVER(data, system):knowledge-structures*
DESIGN { Model Classification, Model Building, Model Validation }	-	•	-	•	•	-	-	•	•	•	•	-	-	*RECOMMEND(base system, required system): change-actions; FORMULATE-DESIGN(components, goals, constrains): system-structure; INTERPRET(data, system): system-state-assessment;*
CHOICE { Evaluation, Sensitivity Analysis, Selection }	-	•	-	-	-	-	-	•	-	-	•	-	-	APPLY-MADM(decision-data):decision-result; WHAT-IF(variable-set, original-model):modified-model; GOAL-SEEKING(goal-variable, original-model):modified-model; IDENTIFY-CRITICAL-VARIABLES(variable-set, original-model):critical-variable-set.

of Service Strategy and the Service Strategy Generation process is used. However, we describe the support from a DMSS in Table 8, and from an i-DMSS in Table 9. The description is described as follows: *to assign priority to a set of alternative service offerings regarding their tag customer expected outcomes (e.g. enhance capabilities, increase performance, enhance resources, reduce costs, and reduce risks).* This is a complex decision, given that the correct selection (authorization) of IT services to be deployed (and implicitly the elimination or postponement of some of them) in the portfolio of services defines the overall IT strategy of an organization. Additionally, relevant financial and other organizational resources will be committed to deploy the selected IT services, and a wrong selection of some IT services will cause a waste of valuable resources.

For the case of a DMSS, we consider the following support: (i) in the Intelligence phase where it is required to identify problems and collect data, the decisional services of ALERT and FIND can help to track large databases automatically (ALERT) or manually (FIND) for cues on patterns of recurrent expected benefits or lack thereof manifested by users in service reports; such data are useful to formulate an adequate scale of customer expected benefits from the planned IT services, and with it to evaluate better the planned portfolio of IT services; (ii) nothing is suggested for the Design phase; (iii) in the Choice phase the decisional services of APPLY-MADM, WHAT-IF, GOAL-SEEKING, and IDENTIFY-CRITICAL-VARIABLES; here numerical pre-defined models (e.g. multi-attribute decision making models) and MADM algorithms are used; APPLY-MADM can provide several mechanisms to rate the alternative service offerings regarding the agreed scale of benefits; WHAT-IF and IDENTIFY-CRITICAL-VARIABLES can help to create awareness in the decision maker on the effects on output variables on minor changes input or intermediate variables;

and finally GOAL-SEEKING can help to identify targets to be pursued on a specific input variable in order to get an expected target in an output variable.

For the case of an intelligent DMSS, we consider the following support: (i) in the Intelligence phase the DISCOVER service is added to the decisional services of ALERT and FIND; also they are added the capabilities of multi-dimensional databases and fuzzy-base SQL processing mechanisms; with such intelligent additions large multi-dimensional databases can be tracked for discovering hidden complex patterns on data of service reports; (ii) in the Design phase with an i-DMSS, in contrast with a DMSS, support can be provided to elaborate the decision Model Scheme; the services of RECOMMEND, FORMULATE-DESIGN and INTERPRET are based on numerical models, knowledge and distributed knowledge bases; expert systems or knowledge-based systems can be used for semi-structured and ill-structured problem-solving situations to elaborate an adequate decisional Model Scheme; (iii) in the Choice phase no additions are suggested as an addition to DMSS.

Based on these illustrations, we propose that both DMSS and i-DMSS are specialized computer-based systems that can and should be used to support IT service decisions. In this chapter, we have reviewed the foundations of a well-known IT service process model (e.g. ITIL v3) as well as the foundations of the decision-making process and DMSS. We then elaborated typical decisions pursued in each IT service phase and process. We continued with two illustrative cases (one DMSS and the other an i-DMSS) to describe how such systems can support a specific IT service decision. In both cases we described the specific architectural and decisional service capabilities that both types of systems can provide.

5 Conclusions

IT Service Management is an important application area in the IT discipline in this decade. Several process frameworks have been reported with international scope. Business and public organizations are interested in deploying IT Service Management to increase IT business value. As with other managerial processes, IT Service Management includes simple, moderate and complex decisions. On the other hand, specialized computer-based systems called Decision-Making Support Systems (DMSS) and intelligent Decision-Making Support Systems (i-DMSS) have been designed and used to support the phases and steps of decision-making processes in several domains. Such systems have benefited decisional processes and outcomes such as improved organizational performance, improved decision quality, improved communication, enhanced mental models, amplified analytical skills of decision-makers, and reduced decision time.

However, DMSS and i-DMSS have seldom been reported in the IT Service Management application domain. In this chapter, we have presented a case for bringing research in these two areas together. To do so, we have shown an i-DMSS framework that offers a useful conceptual tool to analyze the support capabilities of DMSS and i-DMSS. In this framework, three dimensions account for the functional and structural capabilities required for DMSS, with the scale increasing from basic and simple representation, interface and processing to intelligent-based capabilities. Using these dimensions and the components of a DMSS, together with the structure of a generic Decision-Making Process, we have reported two cases to show how a DMSS and an i-DMSS can support a complex decision in IT Service Management. We have also detailed several examples of moderate and complex decisions for this application domain.

We claim that we have established a theoretical basis and provided several research directions to foster joint theoretical and applied research on DMSS, i-DMSS and IT Service Management. Decision-making theories can help IT Service Management to focus on relevant decision tasks or decision problems. DMSS embed decision-making theory within a computer-based system to help the decision maker. AI-based techniques can extend DMSS by incorporating more complex representations for data, information and knowledge models and more intelligent processing algorithms than traditional processing capabilities. Hence, we propose that IT Service Management will benefit from research that considers that application of DMSS and i-DMSS to improve decision making in this domain.

References

1. Lusch, R., Vargo, S. (eds.): The Service-Dominant Logic of Marketing: Dialog, Debate, and Directions. M.E. Sharpe, New York (2006)
2. Sheehan, J.: Understanding service sector and innovation. Commun. ACM **49**(7), 43–47 (2006)
3. Chesbrough, H., Spohrer, J.: A research manifesto for services sciences. Commun. ACM **49**(7), 35–40 (2006)
4. Quinn, J.: Intelligent Enterprise. The Free Press, New York (1992)
5. Spohrer, J., Maglio, P., Bailey, J., Gruhl, D.: Steps toward a science of service systems. IEEE Comput. **40**(1), 71–77 (2007)
6. Mora, M., Raisinghani, M., O'Connor, R., Gelman, O.: Toward an integrated conceptualization of the service and service system concepts: a systems approach. IJIISS. **1**(2), 36–57 (2009)
7. Sage, A., Cuppan, D.: On the systems engineering and management of systems of systems and federations of systems. Inf. Knowl. Syst. Manag. **2**, 325–345 (2001)
8. OECD: highlights of the OECD information technology outlook (2004). http://www.oecd.org . Accessed 10 July 2007
9. Mora, M., Gelman, O., Frank, M., Cervantes, F., Forgionne, G.: Toward an interdisciplinary engineering and management of complex IT-intensive organizational systems: a systems view. IJITSA **1**(1), 1–24 (2008)
10. Olson, M., Chervany, N.: The relationship between organizational characteristics and the structure of the information service function. MIS Q. **4**(2), 57–68 (1980)
11. OGC: The Official Introduction to the ITIL Service Lifecycle. TSO, London (2007)
12. ISO: ISO/IEC 20000-1: Information Technology – Service Management Part 1: Specification. ISO, Geneva (2005)
13. ISO: ISO/IEC 20000-1: Information Technology – Service Management Part 2: Code of practice. ISO, Geneva (2005)

14. van Bon, J., et al.: Foundations of IT Service Management based on ITIL v3. Van Haren Publishing, San Antonio (2007)
15. SEI: CMMI® for Services Version 1.3. Software Engineering Institute, Pittsburgh (2010)
16. EMA: IBM Tivoli Unified Process (ITUP): Connecting the Dots. Business Report. Enterprise Management Associates (EMA), Boulder (2006)
17. Microsoft: MOF Executive Overview version 4.0. Internet document: www.microsoft.com/mof4 (2008)
18. Johnson, M., Hately, A., Miller, B., Orr, R.: Evolving standards for IT service management. IBM Syst. J. **46**(3), 583–597 (2007)
19. Cater-Steel, A., Toleman, M.: Transforming IT service management—the ITIL impact. In: Proceedings of the 17th Australasian Conference on Information Systems, Adelaide, Australia 2006
20. Huber, G.: Managerial Decision Making. Scott, Foresman and Co., New York (1980)
21. Mora, M., Forgionne, G., Cervantes, F., Garrido, L., Gupta, J.N.D., Gelman, O.: Toward a comprehensive framework for the design and evaluation of intelligent decision-making support systems (i-DMSS). J. Decis. Syst. **14**(33), 321–344 (2005)
22. Forgionne, G., Mora, M., Gupta, J., Gelman, O.: Decision-making support systems. In: Encyclopedia of Information Science and Technology, Idea Group, Hershey (2005)
23. Forgionne, G.: Decision Technology Systems: a vehicle to consolidate decision making support. Inf. Process. Manage. **27**(6), 679–697 (1991)
24. Simon, H.: Administrative Behavior: A Study of Decision-Making process in Administrative Organizations. Free Press, New York (1997)
25. Turban, E., Aronson, J.: Decision Support Systems and Intelligent Systems. Prentice-Hall, Upper Saddle River (1998)
26. Phillips-Wren, G., Mora, M., Forgionne, G., Gupta, J.: An integrative evaluation framework for intelligent decision support systems. Eur. J. Oper. Res. **195**(3), 642–652 (2009)
27. Sprague, R.: A framework for the development of decision support systems. MIS Q. **4**(4), 1–26 (1980)
28. Elam, J., Konsynski, B.: Using artificial intelligence techniques to enhance the capabilities of model management systems. Decis. Sci. **18**, 487–501 (1987)
29. Gray, P., Watson, H.: The new DSS: data warehouses, OLAP, MDD and KDD. In: Proceedings of the AMCIS Conference 1996, Phoenix (1996)

Printed in the United States
By Bookmasters

Printed in the United States
By Bookmasters